RADIATION PROCESSES IN CRYSTAL SOLID SOLUTIONS

Author:

Gennadi Gladyshev

St. Petersburg State Technological Institute
Technical University
Russian Federation

eBooks End User License Agreement

CONTENTS

FOREWORD

Interest in the effect of irradiation on the properties of constructional materials was appeared far back with the development of the first nuclear reactors. Since then, the experimental and theoretical studies of radiation destruction processes of pure and doped solid solutions began to develop intensively. As we know the main structural materials used in nuclear reactors, in the irradiation facilities and space exploration are metal alloys. Despite the relative simplicity of the crystal structure in metals the complex processes of formation and interaction of different types of defects under high-energy irradiation take place. Of the other substances such as, alkali metal halides have similar close-paced structure. However, the radiation resistance of metals and alkali halides is very different. The represented eBook provides a comprehensive overview of processes in these materials under the influence of radiation. A detailed description of the mechanisms of radiation defect formation and their evolution in alloys and doped alkali halides gives a clear picture of the radiation-stimulated processes and phenomena in these materials, and allows us to understand the reason for their different radiation resistance.

The eBook contributes to the baggage of scientific knowledge about the effects of radiation on solid materials and represents great interest not only for the specialists in the field of radiation physics and radiation material, but may also be useful for students and teachers of higher educational institutions specializing in this area and related fields.

O. Yu. Begak
Prof., Dr.
Chief Scientist
Institute of Metrology of the DI Mendeleyev,
St. Petersburg, Russian Federation

PREFACE

In the eBook the processes occurring in solid solutions on the basis of metals and alkaline halide under the high-energy radiation are discussed. Metals and alkali halides have a similar close-packed crystalline structure but strongly differ in their radiation resistance. Metals and alloys are radiation-resistant materials, and alkali halides are rather sensitive to radiation. Despite such differences many processes occurring during irradiation are carried out by the identical mechanisms in both types of materials.

The most important result of irradiation is the generation of super-equilibrium concentration of point defects which can be several orders of magnitude greater than the thermal equilibrium concentration. The degree of super saturation of point defects is the main driving force of radiation-stimulated processes in solid solutions. The migration of radiation defects and solute atoms, their interaction among themselves and with the structural damage of the crystal lattice determine the entire spectrum of radiation-stimulated processes. We consider the basic processes and phenomena in crystalline solid solutions under irradiation. These include radiation-induced diffusion, segregation and decomposition of the solid solution, which significantly alter the structure of the material, its physical, mechanical and other properties. Some important radiation induced phenomena, such as radiation-induced swelling, void formation, growth, embrittlement, creep, and the behavior of hydrogen, deuterium and helium in irradiated materials require separate reviews.

Gennadi Gladyshev
St. Petersburg State Technological Institute
Technical University
Russian Federation

2

CHAPTER 1

Kinds and Properties of Radiation: Lattice Defects in Metals and Alkali Halides

Abstract: The main kinds, properties and radiation units of measure are considered. The data on the path of the microparticles in metals and air are presented. The crystal structure of the metals and alkali halides is considered. The models of various kinds of structural defects (point, linear, volume) in crystal lattice are discussed.

Keywords: Radiation, units of radiation, path of the microparticles, crystal lattice, metals, alkali halide, lattice defects.

1.1. KINDS AND PROPERTIES OF RADIATION

Radiation is a flux of micro-particles or electromagnetic waves, which penetrate into the material and cause the damage in its electronic and/or spatial structure. The radiation energy is in the range from a several 0.001 eV to tens (or more) MeV (1 eV = $1.6 \cdot 10^{-19}$ J). Unlike other types of effects (thermal, electric, magnetic, mechanical) energy of the radiation flux is not distributed continuously in the entire volume of the irradiated object, but is allocated locally in some small areas in big portions. The specifics of such exposure lead to the fact that a relatively small integral absorbed energy may cause strong electronic and structural changes in the crystal lattice and, consequently, a marked change in the optical, electrical, mechanical and other properties of crystals. Radiation includes:

- X-rays are electromagnetic waves in a range of wavelengths 10^{-8}-10^{-10} m.

- γ-rays are electromagnetic waves with a wavelengths above 10^{-10} m.

- β-radiation consists of high-energy electron (energy > 1 keV).

- Neutrons. Thermal neutrons (energy $5 \cdot 10^{-3}$-0.5 eV), fast neutrons (energy > 0.1 MeV).

- α-particle consists of 2 neutrons and 2 protons (helium nuclei).

- Ion flux.

The sources of radiation are the reactor emission (thermal and fast neutrons, gamma rays), the particle accelerators (electron, proton, ion), the isotope plant (for example, β-radiation ^{90}Sr-^{90}Y, energy beta particles 0.546 and 2.27 MeV, γ-radiation ^{137}Cs, energy gamma-rays 0.66 MeV, γ-radiation ^{60}Co, γ-ray energy 1.17 and 1.33 MeV, 14 MeV neutrons, alpha particles), X-ray units (electromagnetic radiation with energies of several tens of keV). Power of isotope sources is significantly less than that of electrophysical machines. Typically, power of stationary isotopic emitters does not exceed hundreds of watts.

The efficiency of the interaction between the radiation flux and matter depends on the mass, charge and energy of incident particles. The greater the mass and charge, the stronger the interaction. At the same time, the increase of kinetic energy of the particles reduces the efficiency of interaction. The main effect of the primary interaction of accelerated electrons, gamma and X-ray radiation with matter is ionization. Therefore, these types of radiations are also called ***ionizing radiation***. Quantum energy of ionizing radiation lies in limits from several eV to units MeV. Ionizing radiation, taking into account the weakening over the depth, loses its energy relatively evenly over the volume. UV light is also included in ionizing radiation. In opaque media, it is completely absorbed in thin surface layers. Protons and ions also ionize the matter strongly. However, along with the ionization, these kinds of radiations cause significant damage in the crystal structure (perhaps even mixing, micro-melting). Such radiation is called the degradated or cascade-forming.

Irradiation of microparticles and electromagnetic radiation with energies above 10 MeV and neutrons with any energy can lead to nuclear reactions. The irradiated samples may become radioactive under the action of this type of radiation.

An important characteristic of the interaction of radiation with matter is the linear transmission of energy (dE/dx). Its value depends on the energy, charge and mass of the incident particle and lies in the range from 10^{-4} to 10^5 eV nm^{-1}. Such a large range of variation in dE/dx assumes different mechanisms and consequences of irradiation. For 1 MeV electrons $dE/dx \approx 1$ MeV mm^{-1}. On following example, it is possible to judge about various efficiency levels of radiation effect of various kinds. One 10 keV Cu$^+$ ion transfers about the same energy to the primary knocked atom as does a 1 MeV neutron.

The flux weakening of ionizing radiation, and hence the transfer of radiation energy to the absorbing material is approximately exponentially:

$$I = I_0 \exp(-\mu x),$$ (1.1)

where I_0 is the initial value of the flux intensity, μ is the absorption constant, x is the penetration depth. Despite the continuity of the exponential absorption the pathes of various types of radiation particles are indicated. The higher the radiation energy and the smaller electron density of the medium, the greater the path. Paths of some particles in metals and in the air are presented in Tables **1.1** and **1.2**.

Table 1.1. Path of Electron, ^{4}He^{2+}, ^{12}C$^+$ and ^{20}Ne$^+$ in Al and Ni

Metal	β-particle		α-particle		^{12}C$^+$		^{20}Ne$^+$	
	1 MeV	4 MeV	1 MeV	4 MeV	1 MeV	4 MeV	1 MeV	4 MeV
Path in Al, m	1.5 10^{-3}	7.5 10^{-3}	2.2 10^{-5}	1.3 10^{-4}	1.0 10^{-5}	5.3 10^{-5}	9.3 10^{-6}	4.0 10^{-5}
Path in Ni, m	4.5 10^{-4}	2.3 10^{-3}	9.9 10^{-6}	5.3 10^{-5}	4.8 10^{-6}	2.2 10^{-5}	4.4 10^{-6}	1.7 10^{-5}

Table 1.2. Path of 1 MeV α-Particle, Electron, Proton, γ–ray and Neutron Range in Air

Radiation	α-particle	Proton	Electron	Neutron	γ–ray
Path, m	5.6 10^{-3}	1.8 10^{-2}	3.2	390	820

A completely different dependence of the energy loss with depth is observed during irradiation with heavy ions. In case of ion irradiation, the intensity of interaction, and the concentration of structural defects increase with the depth of penetration and reach a maximum at the end of the ion path, and then decrease. Fig. (**1.1**) shows the distribution profile for 40 keV He$^+$ ions in the depth of the surface and the distribution of stored energy in the form of radiation defects [1]. Fig. (**1.2**) shows normalized rates of point defect production in Ni by various kinds of radiation (1: fast neutrons or high-energy electrons, 2:3 MeV Ni$^+$, and 3: 400 keV protons) [2].

Basu *et al.* [3] investigated the interaction of α-particles with single crystal NaCl. It was found that the path of α-particle (28 MeV) along direction <110> is shorter than along direction <100> and the path in doped crystal is lower than in pure crystal.

The characteristics of ionizing radiation are the ***exposure dose*** and ***absorbed dose***. Exposure dose (D_e) is the ratio of magnitude of the one sign charge (Q), occurring in dry air by the passage of ionizing radiation to the air mass (m) in which this charge is created:

$$D_e = \frac{Q}{m}.$$ (1.2)

In the SI system the exposure dose is measured in the C kg^{-1}. Common unit of D_e is the roentgen (R). One roentgen corresponds to charge $2.08 \cdot 10^9$ ion pairs in 1 cm^3 dry air at 0 °C and a pressure of 101.3 kPa. 1 R $= 2.58 \cdot 10^{-4}$ C kg^{-1}.

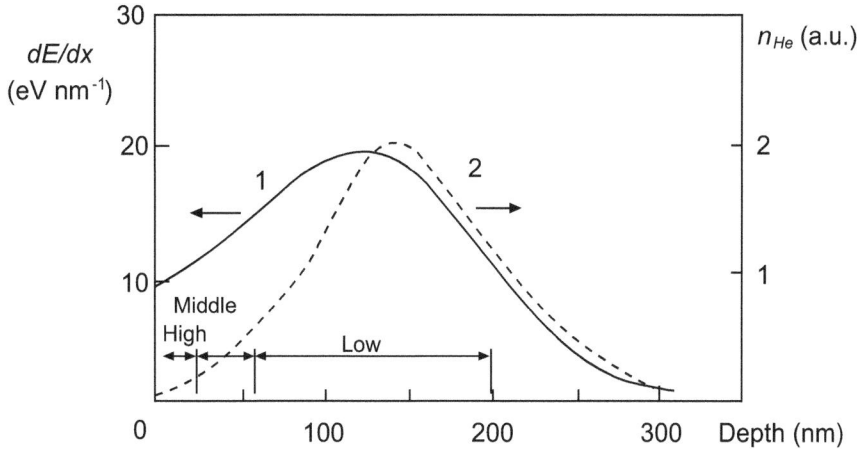

Fig. (1.1). Range and energy deposition distribution as a function of penetration depth for 40 keV He$^+$ ions in Fe-B alloy. High, Middle and Low in the figure indicate the depth ranges in DCEMS measurements. 1: dE/dx (eV nm^{-1}), 2: concentration of He atoms (a.u.) [1].

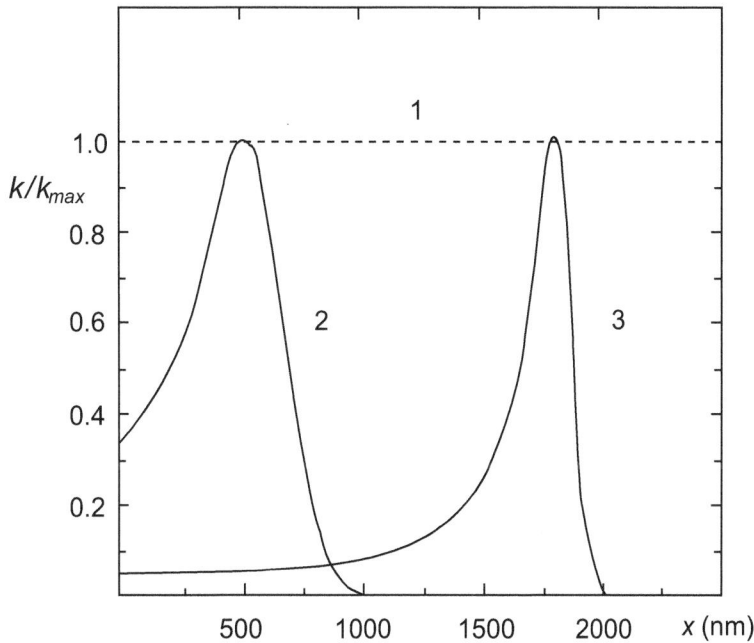

Fig. (1.2). Normalized rates of point-defect production in Ni by different irradiation. 1: fast neutrons or energetic electrons; 2: 3 MeV Ni$^+$ ions; 3: 400 keV protons [2].

Absorbed dose (D) is the amount of radiation energy (E) imparted to unit mass (m) of absorbing substance:

$$D = \frac{E}{m}.$$ (1.3)

The SI unit of absorbed dose is Gray (Gy). 1Gy = 1 J kg^{-1}. Common unit of absorbed dose, used previously, is rad. 1 rad $= 10^{-2}$ Gy. Dose rate is measured in Gy s^{-1}. It should be noted that for many solids 1 R equal $\approx$ 1 rad (≈ 0.01 Gy). In addition, 1 Gy corresponds to approximately 10^{11} reactor neutrons with γ-radiation.

To characterize the radiation damage of metals and alloys there is a widely used unit of measurement known as dpa (number of displacements per atom). The 1 Gy s^{-1} corresponds to approximately 3 10^{-14} dpa s^{-1} for γ-radiation of ^{60}Co. Table **1.3** presents the approximate values of the defect formation rate in the units of dpa for some microparticles with the flux 10^{20} m^{-2} s^{-1}. The approximately equivalent values of absorbed dose in dpa for the flux 10^{22} m^{-2} are shown in Table **1.4**. The data presented in Tables **1.3** and **1.4** was obtained by averaging multiple values from different sources.

Table 1.3. Defect-Production Rate dpa for the Radiation Flux 10^{20} m^{-2} s^{-1}

Material	Radiation	Energy, MeV	Production rate, dpa s^{-1}
Al	electrons	1	6.0 10^{-5}
Ni	electrons	0.65	2.2 10^{-6}
Ni	electrons	2	1.6 10^{-5}
Ni	electrons	3	1.3 10^{-5}
Fe	electrons	1	5.2 10^{-7}
Fe	electrons	2.5	3.3 10^{-6}
Fe	electrons	3	5.6 10^{-6}
Al	n^{o}	fast	8.3 10^{-5}
Au	n^{o}	fast	2.0 10^{-5}
Cu	Cu^{+}	0.300	17.2
Ag	P^{+}	0.270	1.3 10^{-5}

Table 1.4. Dose dpa for Radiation Fluens 10^{22} m^{-2}

Material	Radiation	Energy, MeV	Dose, dpa
Ni	electrons	0.65	2.2 10^{-6}
Ni	electrons	2	1.6 10^{-5}
Ni	electrons	3	3.0 10^{-4}
Fe	electrons	5	4.0 10^{-5}
Fe	n^{o}	fast	8.8 10^{-4}
Al	n^{o}	fast	8.3 10^{-5}
Au	n^{o}	fast	2.0 10^{-5}
Ag	proton	0.270	1.3 10^{-5}
Ni	proton	4	0.05
Fe	Ar^{+}	92	590
Ti	Ar^{+}	2.1	625
Fe	C^{2+}	20	170

1.2. STRUCTURE AND DEFECTS OF CRYSTAL LATTICE

1.2.1. Structure of Crystal Lattice

The objects of the present study are metal alloys and solid solutions based on alkali metal halides. Most metals and alkali-halide crystals (AHCs) have face-centered cubic (fcc) or body-centered cubic (bcc) lattice, and some metals have hexagonal close packing (hcp). Unlike metals, lattice of AHCs is ionic. Most AHCs have a structure of NaCl type (see Fig. (**1.3a**)), which is a combination of cation and anion fcc sublattices. The ions of opposite sign alternate with each other along crystallographic direction <100>. Only crystals CsCl, CsBr and CsI (first at a temperature below 445 °C) have a structure such as CsCl (see Fig. (**1.3b**)). In this case, the simple cubic lattice of cations and anions is located relative to each other so that each ion of one sign should be placed in the center of the cube of the opposite sign sublattice.

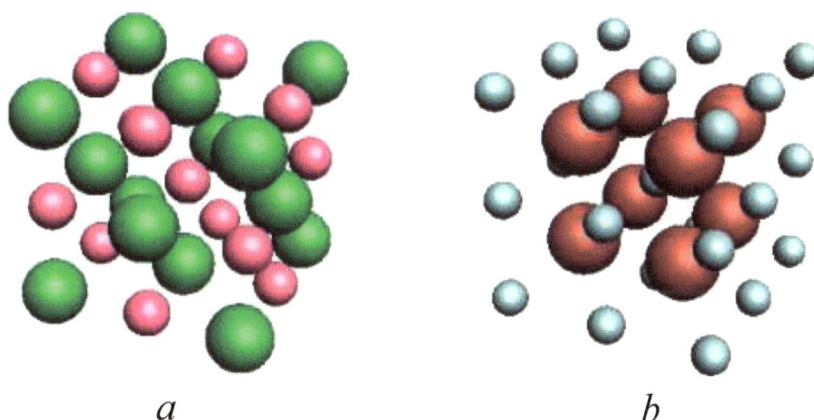

a   *b*

Fig. (1.3). Structure of NaCl (**a**) and CsCl (**b**) type.

At a similar spatial structure, the metals and ionic crystals have a different electronic structure. The first are good electric conductors, and the second are insulators. The difference of electronic structure determines its different radiation resistance. Metals and alloys possess extremely high radiation stability and are the main structural materials exposed to radiation. On the contrary, AHCs refer to the materials with the lowest radiation resistance.

1.2.2. Crystal Lattice Defects

All real crystals contain some irregularities in periodicity of the lattice, called lattice defects. It can distinguish the following groups of defects:

- Point defects (vacancies, interstitials, impurity atoms, small clusters of point defects);

- Linear defects (dislocations);

- Planar defects (sample surface, interfacial surface, grain boundaries);

- Volume defects (voids, impurity precipitates).

Consider the defects that play the most important role in radiation processes.

Point Defects

Simplest lattice defect is a ***vacancy***. Vacancy is the absence of atom in the crystal lattice site. This defect is called a ***Schottky defect*** which in metals causes a deformation of vlume $\Delta V_v \approx - 02\text{-}0.5\ \Omega$ (where Ω is the atomic volume). For example, $\Delta V_v = - 0.4\ \Omega$ and $- 0.22\ \Omega$ for copper [4] and nickel [5], respectively. In ionic crystals, each vacancy with respect to the lattice has a charge which is equal in magnitude and opposite in sign to the ion removed from the site. Therefore, to ensure electrical neutrality of the AHC the Schottky defects should be formed in pairs simultaneously in the cation and anion lattices, forming a $V_c^- - V_a^+$ dipole. According to [6] relaxation of the nearest area of the anion vacancy in NaCl is 40-50 %. However, because of the fact that the formation energy of the cation and anion vacancies is different, their equilibrium concentration will be different. The excess charge in this case is distributed on the crystal surface. This surface charge is called the Debye layer. Vacancies are thermodynamically equilibrium defects, the formation energy (E_v) of which in metals and alkali halides is approximately 1 eV = 1.6 10^{-19} J. The equilibrium single vacancy concentration is determined by thermal fluctuations and is given by expression [7]:

$$n_V = n_0 \exp\left(-\frac{E_V}{kT}\right),$$

(1.4)

where $n_0 = \exp(S/k)$ is the entropy term, k is Boltzmann's constant, and T is the absolute temperature. For example, for copper at room temperature $n_v = 10^{-22}$([1]). Table **1.5** presents the values of S/k, E_v and the vacancy concentration at the melting point ($n_v(T_m)$). The formation energy of Schottky defect (per vacancy pair) in AHCs is in the range 1-3 eV. For example, for KCl and CsCl it is 2.49 eV [8] and 2.6 eV [9], respectively.

Table 1.5. The Values of S/k, E_v [7], the Vacancy Concentration at the Melting Temperature n_v (T_m) for Some Metals

	Al	Au	Cd	Cu	W
S/k	0.7	0.7	0.4	2.4	2.0
E_v, eV	0.66	0.94	0.41	1.27	3.6
$n_v(T_m)$	9.4 10^{-4} [10]	7.2 10^{-4} [11]	5.0 10^{-4} [12]	2.0 10^{-4} [13]	1.0 10^{-4} [14]

In real crystals at room temperature, the vacancy concentration is always higher than thermodynamic equilibrium concentration, defined by Eq. (1.4). This is due to the technology of production. Metals and their alloys, AHCs and solid solutions based on them are mainly produced throught the process of melting. Upon cooling, even for relatively long, most of the vacancies due to the low mobility did not have time to get out on crystal surface to reach equilibrium concentration. Another reason for the high content of vacancies in alkali halides may be the presence of multivalence impurities.

An important characteristic for point defects is their mobility. In most metals, the visible mobility vacancies begin to appear when temperatures approximately above 200 K. According to Hautojarvi *et al.* [15] in iron the vacancies begin to move at 160 K. For most metals the energy of vacancy migration (E_{mv}) is less than 1 eV. However, for example, according to [16] in tungsten E_{mv} is equal to 1.7 eV. The relation $E_{mv} = 6.5kT_m$ can be used to evaluate the activation energy of vacancy migration in metals [17]. In some cases, it is possible to observe the dependence of the vacancy migration energy on impurity concentration. For examples, in Ag-8.14at.%Zn $E_{mv} = 0.64$ eV, in Ag-30at.%Zn $E_{mv} = 0.60$ eV [18]. The vacancy mobility begins to affect the radiation processes in AHCs for the cation vacancies at $T > 250$ K and for the anion at $T > 350$ K. This is due to the difference between migration energy of cation (E_{mvc}) and anion vacancy (E_{mva}). For example, in NaCl $E_{mvc} = 0.85$ eV, and $E_{mva} = 1.12$ eV, in NaBr $E_{mvc} = 0.79$ eV, and $E_{mva} = 1.25$ eV [19]. A relation $E_{mva} = 1.5\text{-}2\ E_{mvc}$ will be correct for the majority of AHCs.

Interstitial atoms (like vacancies) are thermodynamically equilibrium defects. At the same time the any appreciable interstitial concentration can be only at temperatures close to melting since enthalpy of interstitial formation is relatively large. For examples, enthalpyes of interstitial formation in Al, Fe and Pt are 3.2 eV [18], 3.9 eV and 5.0 eV [20], respectively. The stable self-interstitial configuration in both the metals and alkali halides is well established. In metals they exist in the form of dumbbells, lying in the one lattice site [20] (see Fig. (**1.4**)). The nearest area of crystal lattice deforms on the value $\Delta V_i \cong +1.5\ \Omega$ (for example, for copper $\Delta V_i = 1.45\ \Omega$ [4], for nickel $\Delta V_i = 1.7\ \Omega$ [21]) and energy is expended approximately to 5 eV. The dumbbell axis is located along the

---

[1]Here and below the concentration given in mole fraction, except for special cases.

direction <100> in fcc lattices and along the <110> in bcc. Mobility of interstitial atoms is extremely high. Interstitials lose their mobility at temperatures below 50 K. The activation energy of interstitial migration (E_{mi}) is equal 0.1-0.5 eV or lower. For example in Al-Mg alloy the migration energy of free interstitials is equal 0.125 eV [22]. According to Schule [18] the migration energy of self-interstitial in the alloy Ag-Zn depends on the impurity concentration. For Ag-8.14at.%Zn E_{mi} = 0.46 eV, while for Ag-30at.%Zn E_{mi} = 0.41 eV. To estimate the activation energy of interstitial migration in metals, we can use the relation E_{mi} = 0.1-0.2kT_m (T_m is melting temperature) [17].

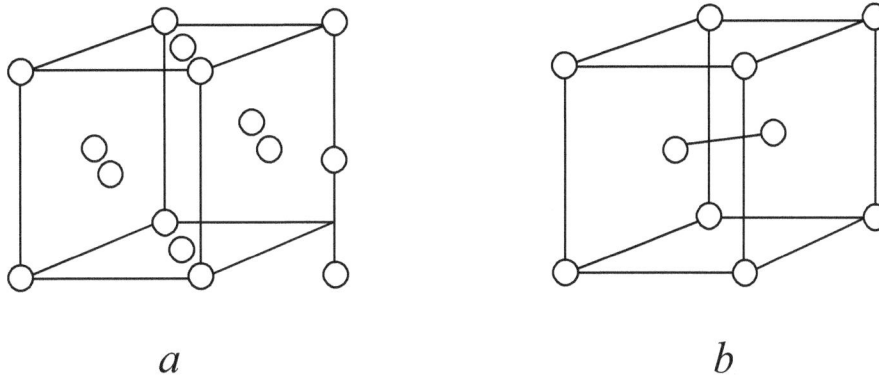

a b

Fig. (1.4). Structure of interstitials in fcc (**a**) and bcc (**b**) metals.

In metals, the least stable configuration of interstitial is *crowdion*. Crowdion is a linear defect, which includes n +1 atoms located in the n lattice sites. The crowdion motion is carried by the most densely packed directions.

When an atom passes from a site to the interstitial position then appears the pair interstitial atom-vacancy. This pair is called a *Frenkel defect*. In ionic crystals, including also AHC, the interstitial position can occupy both cations and anions. Here, just as in metals, the configuration of interstitial defects is a dumbbell in one lattice site. In fcc AHC the dumbbell directs along the axis <110>, in bcc crystals along the <111>. In metals as well as in AHC the interstitial defects are immobile only at very low temperatures. The formation energy of Frenkel defect in AHC for the anion and cation sublattices are different. Their values for some alkali halides are presented in Table **1.6**.

Table 1.6. Formation Energy of Frenkel Defect in Some AHCs [8]

AHC		KCl	KBr	KI	CsCl	CsBr	RbBr
Formation energy of Frenkel pair, eV	Cation sublattice	3.37	3.27	3.02	4.33 [23]	4.05 [23]	3.13 [23]
	Anion sublattice	4.40	4.52	4.60	4.39 [23]	4.55 [23]	4.10 [23]

Impurity Defects

Since the fcc and bcc lattices are close-packed structures, the greater part of the impurity ions is located on the sites of matrix. The solid solution of such type is the substitution solution. Only impurity of small size can occupy interstitial sites or was shifted from the center of site position. For example, one part of carbon atoms in iron is occupied the lattice sites and others part occupies interstitial position [24]. In AHC because of the presence of two different sublattices, all impurities classified into two impurity types: cationic and anionic impurities. Impurities of metal, nonmetal or acid residues are included in lattice, as a rule, in the form of ions and with valence typical for these subjects. The impurities of alkali metals are included in the form of ions M^+, alkaline earth elements have valence 2+, rare earth elements have valence 2+ or 3+, aluminum and bismuth have valence 3+. Anionic impurities are included with its characteristic charge: OH^-, CN^-, NO_3^-, O^{2-}, SO_4^{2-}, CrO_4^{2-} *etc*. It is possible to note some features of placing anionic impurities. For example, ion NO_2^- in the anionic site performs a one-dimensional rotation around the three-fold axis [25]. Sulphur is included into AHC in several forms: S^-, S_3^-, two types S_2^- (in one anionic site and two anionic sites [26]).

When divalent cations (anions) are dissolved in alkali halide crystals, these ions substitute for alkali (halide) ions in the lattice and charge-compensating positive (negative) ion vacancies are introduced in equal number into the lattice. Therefore, divalent cations settle in lattice with the cation vacancy, forming a $M^{2+}-V_c$ dipole. The effective charge of dipole with respect to the lattice is close to zero. For example, in NaCl-Ca the effective charge of dipole is equal to $4.3 \cdot 10^{-2}$ of electron charge [27]. At the elevated temperatures the dipole will be dissociated and the vacancy becomes free. The dipole binding energy lies in the range 0.3-0.6 eV. In AHCs, the undersize impurity can be displaced from a central position of site. Special optical researches of crystals KCl doped with lithium or sodium have shown that Li^+ ion, having considerably smaller radius (0.068 nm) than K^+ ion radius (0,133 nm), takes place in interstitial space of lattice [28]. At the same time the ion Na^+ (0.097 nm) is located at the site of cation sublattice. Polarization properties of impurity luminescence in NaCl-In testify that ion In^+, having smaller size than Na^+, is also located at interstitial position, with corresponding tetragonal symmetry (instead of trigonal, characteristic for central position) [29].

The most part of the impurity ions is in the form of single atoms only under concentration which is smaller than a solubility limit of impurity at given temperature. Depending on the chemical nature of impurity, on its isomorphism in relation to matrix, the solubility can change from extremely low to full miscibility. In alloys, the some impurity can form a various chemical compounds with atoms of solvent, and the phase diagrams of such alloys are quite difficult. We will consider primarily the effect of radiation on solid solutions with limited solubility.

Dislocations

Dislocations subdivide into two types: edge and screw. The ***edge dislocation*** is a linear defect which arises under the introduction of additional (superfluous) half-plane in a crystal lattice. The internal edge of a half-plane is called the core of the dislocation. Near the dislocation core the crystal lattice undergoes elastic distortions. On the one side of core there is a stretching, and on the other side there is a compression. For edge dislocation the shear vector (Burgers vector) is perpendicular to a dislocation line, and for a ***screw dislocation*** the shear vector is parallel to a dislocation line. Preferential slip planes are the atomic planes of most density, and slip directions are the directions of most density in these planes. In metals with fcc lattice, the slip plane is (111), and slip direction is <110>. In bcc metals, the slip plane is (110), and direction is <111>. In the fcc AHC slip plane is (110) and the slip direction is <110>. Any defect can cause elastic distortions. Therefore, under collision of the defect with a dislocation between them arises the interaction either due to elastic attraction (if the distortions are of different characters) or repulsion (if the distortions are of the same sign). The interaction of dislocations with intrinsic and extrinsic defects will lead to the braking of dislocation, and thus to hardening of the crystal. In ionic crystals, if the defect has a charge, then besides elastic interaction, Coulomb interaction is also possible. In real conditions, the so-called mixed dislocations are often dominated. These dislocations have both a screw and edge components, and the Burgers vector lies at an arbitrary angle to the dislocation line. Quantity of dislocations is characterized by a dislocation density. The dislocation density (ρ_d) is defined as the number of dislocation lines crossing a unit area, or as the total length of dislocations per unit volume. In real crystals $\rho_d = 10^{10}$-10^{14} m^2. In the most perfect crystals (in a well-annealed or filamentary AHC) $\rho_d = 10^6$-10^7 m^2, and in cold-worked metals it can reach 10^{16} m^2.

Movement and interaction of dislocations with each other leads to the formation of vacancies, which are mostly in the form of clusters. It is established that vacancies are generated mainly by screw rather than the edge dislocation [19]. Dislocations also interact with impurity defects. Efficiency of interaction depends on elastic stresses arising around the impurity defect under the embedding of impurity in crystal lattice. For example, Br^- and I^- anions in KCl are isomorphic impurities, and therefore they practically are weak obstacles for motion of dislocations [30]. The real manifestation of the interaction of dislocations with impurity atoms is the so-called "Cottrell clouds" [31], i.e. clusters of impurity ions close to dislocations.

Volume Defects

There is a certain probability of formation of defect associates under any concentration of intrinsic and extrinsic defects. Simultaneously with the aggregation of point defects, the reverse processes take place also. Spontaneous accumulation of aggregates of vacancy or impurity atoms is possible only when the concentration of single defects exceeds its solubility limit at a given temperature. Solubility depends on the dissolution enthalpy. Even homologous impurities do not always form a solid solution with unlimited solubility. For example, an equimolar solid solution of

NaCl-KCl begins to decay at 100 °C [32]. Spontaneous aggregation of impurity in the solid solution is also called the ***decomposition of the solid solution***. Usually, aggregation of defect occurs by the standard chemical kinetics which is well described by a system of differential equations [33].

The aggregates can by of various types. If the impurity aggregate has no interfacial boundary, then this aggregate is ***coherent*** and it is called the ***homogeneous precipitate***. These small homogeneous precipitates are formed at the beginning of solution decomposition in the regular regions of the crystal. The large precipitates can also be coherent. For example, the formation of the so-called Suzuki phase has been found in alkali halides doped with some divalent cations [34]. In the crystal KCl-Pb, the structure of the Suzuki phase is $6KCl \cdot PbCl_2$. It was established [35] that the Suzuki phase is formed only when the impurity size of M^{2+} ion is smaller than that of the alkali metal ion of matrix. Mixed aggregates are formed in crystals doped with two different impurities. For example, in the KCl-Eu,Sm crystal were found the mixed Eu-Sm aggregates [36], and in KCl-Pb,Mn were found the mixed Suzuki phase [37].

In the case when the impurity cluster has a structure which different from the structure of the matrix (not coherent with respect to the matrix), then such impurity cluster is called the ***heterogeneous precipitate***. Incoherent precipitates have a well-defined interfacial boundary. In metals, carbides and various intermetalloids are heterogeneous precipitates. In AHC, the impurity halide phase is heterogeneous precipitate. For example, it was found that the Cs_4PbBr_6 micro-crystals are formed in highly doped CsBr-Pb [38] and K_4MnCl_6 micro-precipitates are formed in KCl-Mn ($n_{Mn} = 5 \cdot 10^{-3}$) [39]. Often heterogeneous precipitates appear on the structural imperfections in the crystal (*e.g.* dislocations).

One of the specific phenomena that taks place in supersaturated solid solutions, is the coalescence. The coalescence is a growth of large aggregates due to the dissolution of the aggregates of small size. In this case, the equilibrium concentration of single atom impurity depends on the aggregate radius according to the Thompson' equation:

$$\ln\frac{c_r}{c_\infty} = \frac{2\sigma\Omega}{rkT} , \qquad\qquad (1.5)$$

where c_r and c_∞ are the equilibrium impurity concentrations for particle with radius r and plane boundary, respectively, σ is the interfacial energy, and Ω is the atomic volume.

CHAPTER 2

Intrinsic Radiation Defects and Their Production Mechanisms

Abstract: Models and mechanisms of intrinsic radiation defect formation in the metals and alkali-halide crystals are considered. The effect of irradiation conditions (temperature, absorbed dose, etc.) on the formation of the primary and secondary defects, fomation of the aggregate radiation defects (voids), and affection of heat treatment on the properties of irradiated metals are discussed herein. It is noted that low radiation resistance of alkali halide crystals is connected to the under-threshold mechanism of defect formation.

Keywords: Radiation, metal, alkali halide, color centers, radiation defects, mechanism of radiation defect formation.

Subject of the crystalline solution radiation physics comprises two major primary phenomena: Change of electronic structure and periodicity of the crystal lattice. These processes, as well as the subsequent process of dissipation of the energy of "hot" electrons and atoms, interaction of the resulting radiation defects between each other and with solid solution components, thermal and radiation-stimulated substance transfer determine the entire wide range of radiation transformations. Such affection lasting for 10^{-12}-10^{-11} s results in the formation of secondary high-energy electron cascades and X-ray photons (**secondary irradiation**). Electron excitations and structural defects occur under impact of such primary and secondary irradiation. Type of electron excitations depends on the initial crystal electron structure. This difference in particular serves as the reason of variable radiation stability for metals and AHC. There are two mechanisms of structural defects formation: **elastic** (threshold) and **non-elastic** (all other). Elastic mechanism of defect formation is the most direct and manifests in all materials.

2.1. METALS

2.1.1. Primary Defect Formation

Valence band and conduction band overlap in metals in compliance with the band model. Thus metal valence electrons are regarded as gas with the energy distribution of the Fermi-Dirac statistic. When metals are exposed to radiation high-energy particles interact with electron gas, with electrons of ions' inner shells and with the crystals lattice ions as with individual objects causing ionization and formation of structural defects (Frenkel pairs). Result of the impact depends on the initial attributes of penetrating radiation. Irradiation by destructive radiation results in consumption of the major part of irradiation energy by displacement cascades, by micro-area formation with high density of associates of vacancies and interstitials. The entire process from atoms formation dissipating energy to level under which defects are not generated any more is called the displacement cascade or collision cascade. According to Kirk and Blewitt [40] around 8000 disarrangement displacements fall on one cascade formed by the primarily knocked-on atom of the energy $E > 20$ keV.

Unlike fast neutrons and ions ionizing irradiation does not cause displacement cascades. Electrons and gamma rays generate isolated Frenkel pairs with more flat distribution in terms of volume. Intense flows of ionizing irradiation result in sudden increase of free electrons which causes increase of internal pressure. If electron flow equals to 10^{26}–10^{27} cm^{-2} s^{-1} internal pressure in metals achieves several kbar [41].

Mechanism of the primary structural defect formation is explored fairly well. Excluding the case of irradiation by ultra-high energy ions, defects formation in metals occurs by **threshold mechanism**, which is based upon classical mechanics laws. From the positions of binary collision energy (E) transmitted to the ion will depend on kinetic energy of incident particle (E_k), particle mass (m) and atom mass (M):

$$E = E_k \frac{4mM}{(m+M)^2}.$$

(2.1)

Relativistic effect shall be taken into account in electron irradiation. In this case the maximum transmitted energy will be as follows:

$$E = E_k \frac{2(E_k + 2m_e c^2)}{Mc^2}.$$ (2.2)

To knockout atom from the node it is necessary that $E > E_d$, where E_d is the threshold energy. E_d value for various metals shall be within the range 7-80 eV, with a commonly accepted average 25 eV. For example, for rare-earth elements Y, Sm and Yb it shall be equal to 14.6, 9.5 and 8.7 eV, accordingly [42]. It is interesting to note that the energy of Frenkel pair thermal formation in metals equals to 3-6 eV.

Evidently the defect formation rate depends on the radiation type, its energy, irradiation flow and temperature. Thus in metals by 100 keV energy of particle, neutron will displace around 10^4 atoms, proton a bit less, electron will displace a few atoms, gamma-quantum - not more than one. The calculations of Terentyev *et al.* [43] made by method of molecular dynamics for the alloy Fe-10%Cr indicate that the highest concentration of radiation defects is achieved within several picoseconds. Due to the high mobility of interstitial defects major part of Frenkel pairs is recombined within the time of 10^{-12}-10^{-11} s approximately and restores the original structure of metal. According to Kiritani [44] 85 % of all interstitials in gold generated by fission neutrons at room temperature are recombined inside mother cascade and another 10 % in neighboring cascades. The remained 5% are captured by dislocations and dislocation loops. Some observation data [45] indicate that under similar irradiation conditions the defect density in metals with FCC structure is higher than in metals with BCC structure. Thus after irradiation by neutrons under the temperature of 150 °C and dose $\varPhi = 2.2 \ 10^{25}$ m^{-2} defect concentration in Ni and Cu (fcc) amounted to $n_d = 2 \ 10^{23}$ and $9 \ 10^{21}$ m^{-3}, whereas in Mo and Fe (bcc) it amounted to $n_d = 10^{21}$ and $8 \ 10^{21}$ m^{-3}, respectively. Meanwhile there are less defects of interstitial type in Ni and Cu than of vacancy type by one or two order respectively. In Mo number of interstitial and vacancy defects is equal and in Fe interstitial type of defect prevails over vacancy.

Steady-state concentration of defects plays the most important role in radiation processes occurring in solid solutions. The time of steady-state concentration establishment for point defects depends mainly on temperature and activity of the sinks. According to Kozlov's data [46] quasi-equilibrium concentration of interstitial atoms in austenitic steels is achieved during $\sim 10^{-6}$ s and of vacancies – during $\sim 10^3$ s when they are irradiated by neutrons ($\phi = 10^{-10}$-10^{-2} dpa s^{-1}) at the temperatures of 373-773 °C. According to other data [47] when nickel is irradiated by 100 keV He$^+$ ions ($\phi = 4 \ 10^{17}$ m^{-2} s^{-1}) at 500 °C the equilibrium concentration of defects is achieved after 4.5 s of irradiation.

2.1.2. Point Defect Clusters

The long irradiation leads to the accumulation of structural defects. Meanwhile the decisive meaning will have the defects supersaturation degree which is the driving force for some radiation-induced phenomena. Brailsford and Bullough [48] suggested that the degree of supersaturation by vacancies should be assessed in the following way:

$$S_v = \frac{K}{k_v^2 D_s},$$ (2.3)

where K is the defect formation rate, k_v^2 is the sink strength for vacancies, D_s is the thermal self-diffusion coefficient. If supersaturation of defects is maintained within a long-time period at the temperature efficient for the vacancy movement, vacancy clusters will be formed. The growth of vacancy complexes results in void formation. For example, void formation in pure iron is observed after irradiation by 700 keV electrons ($\phi = 4.4 \ 10^{-4}$ dpa s^{-1}) at the temperature of 430-490 °C and doses of less than 5 dpa [49]. Vacancy clustering is specifically efficient under heavy ions irradiation. Sekimura *et al.* [50] found that the efficiency of vacancy cluster formation depends on a kind and energy of ions. After irradiation by ions V$^+$ of the energy 200 and 400 keV and dose of $\varPhi = 10^{16}$ m^{-2} at room temperature, vacancy complexes have not been found. At the same time irradiation by ions Au$^+$ with energy exceeding 120 keV causes cluster formation in the size 2-2.5 nm. Irradiation of stainless steel by 1 MeV electrons at 550 °C within several hours causes the void formation [51]. After six hours of irradiation the void density approximates the saturation and achieves the value of $\rho_{void} = 1.12 \ 10^{19}$ m^{-3}. However, after 15 minutes of ageing in the result of 1 hour re-irradiation ρ_{void} increases sharply up to the value of 1.12 10^{20} m^{-3}.

The irradiation of metals and the alloys with voids in initial status results in redistribution of voids in terms of size. With the help of method of low-angle neutron scattering Lebedev *et al.* [52] investigated the initial as well as several years irradiated by reactor radiation (with the dose of fast neutrons 2 10^{25} m^{-2}) alloy samples Al–Mg–Si. Analysis of the obtained results demonstrated that after irradiation the total volume fraction of defects $\approx$ 0.5 % increases by 10 % and the total surface of the defects increases by $\approx$ 40 %. The volume concentration of large dissipating areas of the radius $\approx$ 40−50 nm reduces, meanwhile the share of fractions of the size 5−10 and 20−25 nm increases. This may be explained by fragmentation of large areas.

Dislocation and vacancy loops are the primary volumetric structural violations in metals. Vacancy loop develops into flat void. Spherical shape of the void becomes more advantageous with the growth of the void size. Under definite irradiation conditions and significant supersaturation with vacancies voids' ordering can be observed. The voids start to get arranged along the lines and planes and even form *void lattice* [53]. Void ordering is possible only in the case when void size exceeds some critical value [54]. Dependence of the critical radius ($r_v{}^*$) on void density (ρ_v) for some values of dislocation density (ρ_d) and capture radius of self-interstitial atoms by dislocation (r_d) is represented in Fig. (**2.1**).

Void lattice has the symmetry and orientation similar to the matrix lattice. A necessary condition for void lattice is the high rate of displacement formation exceeding some critical value. Void radiuses have approximate equal sizes and are changed within 1-30 nm for various metals. Void lattice constant exceeds void radius by 3-15 times. When radiation intensity is increased the distance between voids is reduced. As temperature increases the void size and the distance between voids is increasing too. However, heating up to higher temperatures leads to annealing of void lattice. Chen and Ardell [55] observed the void lattice formation in the alloy Ni-Al when it was irradiated by 400 keV ions $N_2{}^+$ (ϕ = 2.8 10^{-3} dpa s^{-1}, $\varPhi$ = 70 dpa) at 500 °C. Void lattice had fcc structure and was parallel to the matrix lattice. The most perfect void lattice was found in the alloy Ni-2wt%Al. No void ordering was observed in pure nickel under the same irradiation conditions.

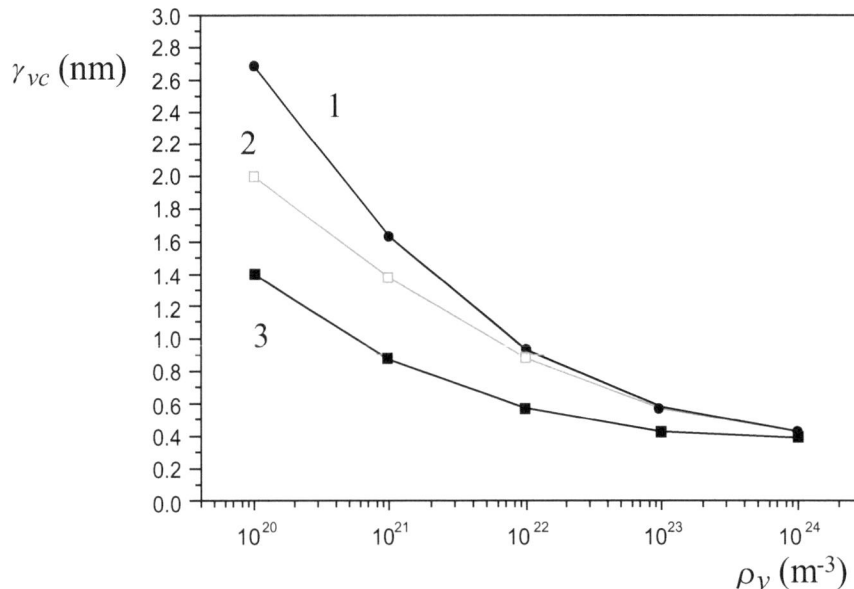

Fig. (**2.1**). Dependence of the critical void radius ($r_v{}^*$) for the onset of void ordering on the number of void density (ρ_v) [54]. 1: ρ_d = 10^{13} m^{-2}, r_d = 10 nm; 2: ρ_d = 10^{14} m^{-2}, r_d = 1 nm; 3: ρ_d = 10^{13} m^{-2}, r_d = 1 nm.

The formation of interstitial complexes may happen in the same effective way as vacancy aggregation. Yang *et al.* [56] investigated impact of high-dosage 1 MeV electron irradiation on pure aluminum. After the irradiation by dose 10 dpa at the temperature of 20-200°C dislocation loops were observed and major part of theirs is of interstitial type. On the contrary voids are mainly formed in nickel and copper during irradiation by electrons at the respective temperature. Irradiation by electrons does not cause swelling, but causes the formation of dislocations and dislocation loops. According to Bae *et al.* [57] the dislocation loop growth was observed and the density and size of dislocation loop increased with dose growth during 1.25 MeV electron irradiation in temperature range from 573 to

873 K. Meanwhile their size and density increased with growth of the dose absorbed. At the same time void formation was not observed up to dose 5.4 dpa.

Rauch *et al.* [58] investigated by method of diffuse scattering the formation efficiency of point defect clusters in copper and aluminum irradiated by fast neutrons at the 4.6 K to a dose of 2.4 10^{22} m^{-2}. It was established that in the irradiated copper after annealing up to 60 K the interstitial loops with the average radius of 1.1 nm are formed. Further annealing to 300 K results in growth of radius to 1.8 nm. Meanwhile 25% of initially formed defects are remained. Vacancy loops of the average size of 1 nm has also been detected at 60 K. In the formation of these loops are involved 20-40 % of the vacancies formed during irradiation. During subsequent heating up to the room temperature the number of vacancies included into the loops and the loop size remains stable enough. Lop formation has not been detected in aluminum. Irradiation of the Zr-2.5wt%Nb alloy by fast neutrons of the dose ~10 dpa at 570 K causes the formation of network dislocation [59].

Intense formation of point defects clusters results in swelling of metals and alloys. Swelling is observed in the zone of intermediate temperatures when the vacancies are quite mobile and concentration of thermal vacancies does not exceed yet the steady-state concentration of radiation vacancies. Evidently the most important factor for void formation is the dose absorbed not the defect formation rate. Sometimes the void formation is observed under very small dose rates. Okita *et al.* [60] discovered that cavity formation in austenitic steels happens already under irradiation by neutrons with $\phi = 8 \ 10^{-7}$ dpa s^{-1} and at the dose 28.8 dpa. Observation data for various alloys demonstrate that the swelling efficiency even increases in some cases with dose rate drop. Such fact, for example was established during irradiation of austenitic steels by neutrons within the range of $\phi = 10^{-4}$-10^{-3} dpa s^{-1} at the temperatures of 300-600 °C [61]. Additional introduction of point defect sinks may result in swelling reduction. Thus deformation of austenitic steel on 20 % after which dislocation density equals to ~10^{15} m^{-2}, reduces several times the formation of vacancy clusters under irradiation by 5 MeV electrons with $\phi = 10^{-4}$-10^{-3} dpa s^{-1} [62].

Dunlop *et al.* [63] investigated radiation damage in alloys Cu$_3$Au, Ni$_3$Fe and in austenitic alloy Fe-Cr-Ni under the flow of ions ^{129}Xe of the ultrahigh energy 3.4 GeV at 4.2 K. If with the increase of incident ion energy the threshold of energy loss per length unit is exceeded 15 keV nm^{-1}, the increase of defect formation rate is observed in amorphous alloys dramatically.

Under irradiation by ionizing radiation major part of the energy absorbed is used for the excitation of electronic subsystem and generation of excessive number of free electrons. Bakay *et al.* [64] demonstrated that due to the small time of electron thermalization, deviation of electron distribution function from the equilibrium state is small even under comparatively high intensities of ionizing radiation and cannot affect significantly macro-physical attributes of metals. Electron excitations are transformed into thermal lattice vibrations in the process of thermalization. The rate of formation and accumulation of structural defects in this case is low. For example [65], in nickel after irradiation by 2 MeV electrons of the fluence 9 10^{21} m^{-2} at the temperatures of 240, 290 and 303 K the electrical resistance is not changed. This indicates to the fact that the concentration of Frenkel defects is less than 2.5 10^{-6}. Metal structures can be operated within tens of years without vivid change of their operational attributes under impact of ionizing irradiation (if these are not super-power fluxes).

2.1.3. Effect of Irradiation on the Electrical Properties

Formation of structural damage leads to a change in the electrical properties of metals. Massey *et al.* [66] calculated the changes of the electronic system potential energy in the point defect proximity. According to the calculations local change of electron density oscillates and reduces at large distances in compliance with the law r^{-3}. Such distortion shall cause additional dissipation of conduction electrons and reduce the effective mean free path of electrons. That is why the resistivity grows with the increase of defect concentration. For pure metals and many alloys the change of resistivity ($\Delta\rho_e$) is proportional to the square root from dose absorbed (Φ) at intermediate temperatures. Such dependence is observed, for example, in copper and some of its alloys at room temperature under irradiation by neutrons up to $\Phi = 3 \ 10^{21}$ m^{-2} [67]. It was established that for one Frenkel defect $\Delta\rho_e$ is equal to 2.0 μΩ cm %$^{-1}$ for copper [4], ~ 60 μΩ cm %$^{-1}$ for alloy Ti-Al [68]. In pure nickel one Frenkel pair changes the resistivity by 97-109 μΩ cm at.%$^{-1}$ [69].

Generation of radiation defects causes perceptible change of thermo-EMF (thermo-electromotive force). Table **2.1** represents results for several metals and alloys, obtained by comparison of EMF values after fast neutron irradiation with non-irradiated samples [70]. In general the total change of thermo-EMF is composed of two components: on the account of radiation defects of structure and on the account of alloying in the result of nuclear reactions. When the temperature is measured with the help of thermocouple the thermal annealing of radiation defects shall also be taken into consideration under irradiation. Due to thermal transformations of radiation defects the induced thermo-EMF will depend on irradiation temperature. At high temperatures when all radiation defects are annealed, the radiation correction into thermo-EMF will be conditioned only by solute nuclear-doped atoms. Platinum-molybdenum thermocouple is quite stable against the reactor neutron irradiation. No influence of fast $1.5 \cdot 10^{25}$ m^{-2} and thermal neutrons $1.5 \cdot 10^{25}$ m^{-2} was detected on indications of thermocouple at the measurement precision error.

Table 2.1. Radiation-Induced Thermo-EMF for Some Metals and Alloys

Metal, Alloy	Radiation	Dose, cm^{-2}	EMF, μV K^{-1}	Reference
W	Deuterons , 13.6 MeV	$2 \cdot 10^{18}$	+0.11	[70]
W	α-particles, 27 MeV	$2 \cdot 10^{18}$	+0.40	[70]
Mo	Protons, 6.8 MeV	$1 \cdot 10^{16}$	-0.017	[70]
Mo	Neutrons, fast	$1.8 \cdot 10^{20}$	-1.04	[71]
Cu	Neutrons, fast	$1.8 \cdot 10^{20}$	-0.36	[71]
Fe	Neutrons, fast	$7.5 \cdot 10^{20}$	-0.14	[71]
Pt	Neutrons, fast	$1.8 \cdot 10^{20}$	-0.08	[71]
Au	Neutrons, fast	$1.4 \cdot 10^{20}$	+0.78	[71]
Chromel	Neutrons, fast	$1 \cdot 10^{20}$	-0.20	[71]
Alumel	Neutrons, fast	$1.2 \cdot 10^{20}$	+0.43	[71]
Constantan	Neutrons, fast	$1.8 \cdot 10^{20}$	+0.19	[71]

Radiation impact results in significant deterioration of superconductive properties of special alloys. Table **2.2** represents the decline rate of critical superconductivity temperature for several alloys.

Table 2.2. Radiation-Induced Reducing of the Critical Temperature of Superconductivity for Some Alloys

Alloy	Radiation	Fluence, m^{-2}	Irradiation temperature	Decrease of critical temperature	Reference
Nb$_3$Sn	Ne^{2+}	$4 \cdot 10^{17}$		3 times	[72]
Nb$_3$Sn	Fast neutron	$1 \cdot 10^{24}$	< 28 K	From 17.9 to 2.5 K	[73]
Nb$_3$Al	Fast neutron	$1.2 \cdot 10^{23}$	413 K	52 %	[74]
V$_2$Zr	Fast neutron	$1 \cdot 10^{24}$	< 28 K	From 7.8 to 2.7 K	[73]

2.1.4. Thermal Processes in the Irradiated Metals

Heating of irradiated alloys results in the annealing of radiation defects and to ρ_e reduction. Dependence of ρ_e on the annealing temperature bears important information about stored radiation defects. Such dependence for pure platinum irradiated by electrons up to the dose equivalent to Frenkel pair concentration $5 \cdot 10^{-5}$ ids represented in Fig. (**2.2**) [75]. Each resistivity reduction is conditioned by annealing of radiation defect of definite type. Significant recovery of resistivity at annealing stage I which is subdivided into some number of sub-stages which are usually called as I_A, I_B, I_C, I_D and I_E is observed at low temperatures. Recombination of proximate Frenkel pairs of various configurations occurs at stages I_A, I_B and I_C. Thermally released interstitial atoms become free and can interact with various types of defects at stages I_D and I_E. Stabilization of not only single interstitial atoms but of paired interstitial defects is possible under low-temperature irradiation. For example, such defects are formed in iron under irradiation by 5 MeV electrons and become movable at annealing stage I_{D1} [76]. At stage II the growth and rebuilding of interstitial clusters and release of interstitial atoms from the complexes of solute-interstitial originated at stage I, occur. Annealing stage III is conditioned by vacancy movement, their agglomeration and recombination with interstitial clusters. Within the temperature area between stage I and III of radiation defect annealing only interstitials are movable in metals. Each metal is characterized by its temperature stages of resistivity recovery. For example, for aluminum stage I appears at 40 K, stage III – over 200 K [77]. Annealing of pure cobalt irradiated by 3 MeV electrons at 8 K also demonstrates three classical annealing stages [78]. Stage I is subdivided into sub-stages 15 K, 33 K, 42 K and 52 K. Stage II has two sub-stages 112 K and 169 K, whereas stage III extends within the temperature area 300-350 K. The type of resistivity recovery curve depends on the availability of this or that solute. Investigation of Al-Mg alloys irradiated by neutrons at 24 K demonstrated that magnesium addition to aluminum results in degradation of recovery in the annealing stages I and II and in increase at the stage III [20].

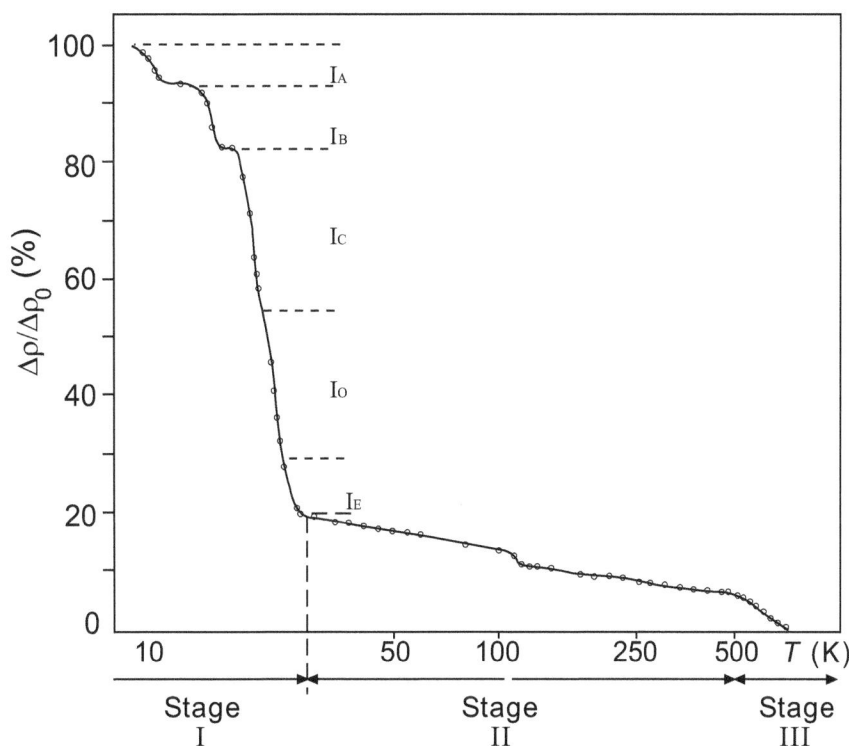

Fig. (2.2). Isochronal recovery of pure Pt irradiated at 4.5 K with 3 MeV electrons to Frenkel pair defects 5 10^{-5} [75].

At the same time there are examples in the literature of ordinary one-stage dependence of resistivity recovery. Such dependence was obtained for pure bismuth and for alloys Bi-Te and Bi-Sn after irradiation by 1 MeV electrons at 20 K to the dose of 4.5 10^{-5} dpa [79]. The only annealing stage at 40 K is conditioned by full recombination of Frenkel pairs due to interstitial movement. At doses exceeding 2 10^{-3} dpa or at electron energy of 2.5 MeV several annealing stages are observed. The annealing stages at 80 and 130 K are ascribed to vacancy and divacancy movement.

Pfeiler and Poerschke [80] have established that two annealing stages of resistivity were observed in the alloy of silver with high concentration of aluminum after irradiation by 2 MeV electrons (ϕ = 2.5 10^{13} cm^{-2} s^{-1}, n_d = 3.5 10^{-6}) and only one stage in the alloys Cu-Al and Cu-Mg. The annealing step at around 80 K is connected to migration of interstitial atoms and steps at 240, 310 and 250 K were caused by vacancy movement in the alloys Ag-Al, Cu-Al and Cu-Mg, respectively. Activation energies of stages for interstitial and vacancy movement coincide with the activation energies of respective defect migration (0.3 and 0.79 eV for the alloy Ag-Al).

Analysis of resistivity recovery, the data obtained by other investigation methods demonstrate that irradiation and the further annealing lead to a significant change of metal and alloy structure. Ehrhard and Schlagheck [81] have established that at the end of annealing stage I (40 K) clusters which include 5-10 interstitials are formed in copper irradiated by electrons. In the course of the further annealing cluster size is increased. At the beginning of stage III (200 K) they include 80 interstitials and at the end of stage III (300 K) the number increases up to 500. Interstitial clusters form loops in planes {111} and {110}. Small size vacancy clusters are also formed at the end of stage III.

The study of radiation damage under very low temperatures is of definite theoretical and practical interest. This is important in the connection with operation of thermonuclear facilities where superconducting-coil electromagnets are used. As Kozlov indicated [82], irradiation of austenitic steel under helium temperatures when thermal mobility of any defects is not available results in the formation of area degraded by radiation and consisting of three areas with different vacancy and interstitial concentrations. Diameter of such area amounts to approximately 20 lattice constant. Central area is enriched with vacancies (about 25 %). The centre is surrounded by an area containing both interstitials and vacancies. The next external area is enriched with interstitials.

A large number of theoretical calculations have been completed to date with the help of different methods in relation to the matter of radiation defect formation and interaction in metals and alloys: kinetic models, the method of molecular dynamics, modeling by Monte Carlo simulation (for examples [83-85]).

2.2. ALKALI HALIDE CRYSTALS

2.2.1. Mechanisms of the Point Defect Formation

Hundreds of articles were devoted to the mechanisms of radiation defect formation in AHC. The most exhaustive and reliable information on the topic is represented in several reviews, for example, in [17, 18]. Unlike metals the valence and conduction bands in AHC are separated by band gap (see Fig. (2.3)). The gap value lays in the range 5.5-13 eV. Such a large magnitude of gap does not permit electrons to transit from the conduction band to the valence band by means of thermal fluctuations. That is why there are no free electrons in AHC under normal conditions. During irradiation by ionizing radiation flows the high-energy electrons generated within a short period of time are exchanged for less energetic and generate free electrons and holes within wide range of energies. In the relaxation process due to strong electron-phonon interaction two thirds of the energy absorbed transforms into vibration energy [86]. The remained part of the energy is used to the formation of excitons and electron-hole pairs, to luminescence, which may take in doped AHC up to 15% of the energy absorbed, to generation of Frenkel pairs in terms of threshold mechanism if energy of incident particles exceeds E_d. Exciton and electron-hole pair formation in AHC is the principle difference in the behavior of alkali halides and metals under irradiation. Namely the formation of low-energy electron excitations and the realization of non-elastic mechanism of structural defect generation serve as the reason for such significant difference of metal and AHC radiation stability.

It is generally accepted now the exciton formation in anionic sublattice is the main factor determining AHC radiation stability. In AHCs, the anionic exciton is an excited halide ion, whose energy level structure is similar to that of the hydrogen atom [87]. Excitons are formed either directly in interaction of secondary electrons with halogen atoms or during recombination of thermalized electrons and holes. The formed free ("hot") exciton has a path length $10-10^4$ of a_0 (a_0 is lattice constant) and relaxing during the time of $10^{-14}-10^{-13}$ s transforms into self-trapped exciton [87]. The self-trapped exciton has the same configuration as the self-trapped hole which has captured electron. Models of the primary and self-trapped excitons and of F and H centers are represented in Fig. (2.4). A path length of thermalized exciton (l_{ex}) prior to its decay has a great importance for radiation processes occurring in AHCs. According to Lushchik N.E. *et al.* [88] l_{ex} is 4 a_0, 20 a_0 and 120 a_0 in KCl, KBr and CsBr, respectively.

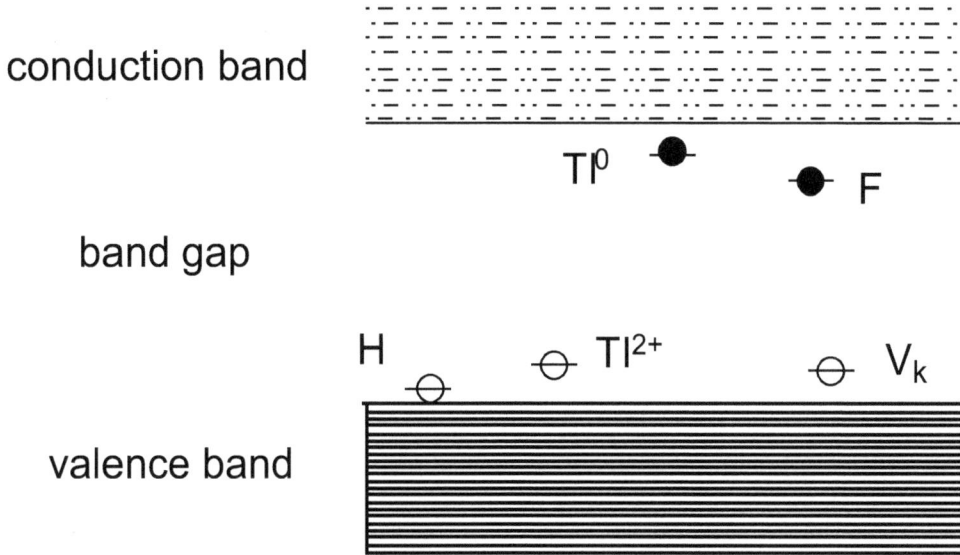

Fig. (2.3). Band structure of the alkali halide crystal.

As far as exciton annihilation energy used for Frenkel pair formation is much less than the threshold energy required for the shock mechanism realization, such mechanism was called **under-threshold**. The most reliable mechanism connected to exciton annihilation is considered the mechanism arising during recombination of electron-hole pairs:

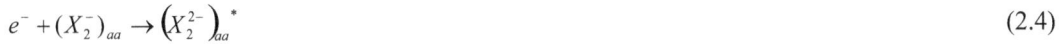

$$e^- + (X_2^-)_{aa} \rightarrow (X_2^{2-})_{aa}^* \tag{2.4}$$

Here e^- is electron, $(X_2^-)_{aa}$ is self-trapped hole (V_k center), $(X_2^{2-})_{aa}^*$ is exciton, superscript denotes the type of lattice site. Special experiments with pulse irradiation demonstrated convincing that the Frenkel defect formation in AHC occurs under nonradiative exciton annihilation within several picoseconds according to the scheme:

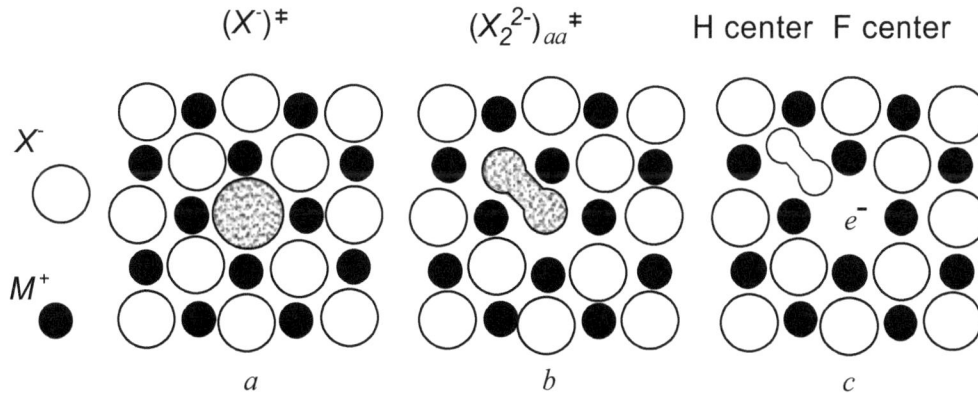

$$(X_2^{2-})_{aa}^* \rightarrow (X_2^-)_a + V_a e^- + X_a^-, \tag{2.5}$$

Fig. (2.4). Radiationless formation mechanism of Frenkel pair in AHC. **a**: non-relaxed ("hot") exciton, **b**: self-trapped exciton, **c**: Frenkel pare (H and F center).

Here $(X_2^-)_a$ is the interstitial halogen ion (H center), $V_a e^-$ is the anion vacancy with electron (F center), X_a^- is the halogen ion in site of anion sublattice. An important factor in the formation of Frenkel pairs is the presence of close-packed crystal directions along which they may displace crowdion interstitials. Hersch and Pooley [89, 90] were the first to suggest the hypothesis on originating of structural defects in regular node of crystal lattice under non-radiative annihilation of excitons. In the result of non-radiative exiton decay the charged Frenkel pairs can appear (I and α centers):

$$\left(X_2^{2-}\right)_{aa}^{*} \rightarrow \left(X_2^{-}\right)_a + V_a + X_a^{-}.$$ (2.6)

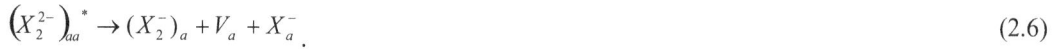

However the probability of Frenkel neutral pair formation is significantly higher than of the charged pair as formation energy of I-α is higher than H-F pair. I centers are thermally less stable than H centers due to their larger size. Migration energy of I center is lower than of H center [91].

To form stable Frenkel pair its components division is required. During irradiation major part of the Frenkel pairs is annihilated and only smaller part of them is stabilized. Only one pair survives from approximately 30 primarily formed F-H pairs. It was established that probability of Frenkel pairs' component division is reduced with lattice energy increase. Frenkel pair formation under direct irradiation of AHC by ultra-violet in exciton absorption band is the important proof of exciton mechanism of defect formation. Tanimura and Itoh demonstrated experimentally transformation of the self-trapped exciton into a pair consisting of interstitial atom and vacancy divided by a distance exceeding two lattice constants [92].

Main processes of radiation defect formation in AHCs occur in anionic sub-lattice. However in some cases cationic sub-lattice may also participate in radiation defect generation. As investigations on CsCl crystals indicated excited crystal areas start to overlap under high density of ionizing radiation (2 10^{24}-2.6 10^{25} eV m^{-3}) [93]. In this case ionizing density of 10^{24} m^{-3} for caesium ions is achieved and conductivity electron is able to recombine with cationic holes (Cs^{2+}) for the time less than 1 ns with cationic exciton formation. Cationic excitons transmit energy effectively to anions with the formation of anionic excitons. Annihilation of anionic excitons results in luminescence or in of Frenkel pair origination. It is also supposed that Frenkel pairs can be created directly in cation sub-lattice [87, 94].

Other under-threshold mechanisms of radiation defect formation in alkali halide were also suggested. Thus Varley [95] suggested mechanism of double ionization, and Seitz [96] suggested the mechanism of exciton localizing on dislocation with the further annihilation and creation of Shottky pair. However it was demonstrated that mechanisms of Varley and Seitz made no significant input to defect formation in AHCs.

During irradiation by electrons or gamma-rays major part of their energy is used for excitation of electronic sub-system not for elastic displacement. At the same time shock mechanism of defect formation is naturally realized under effect of ionizing radiation. Meanwhile minimal gamma-quantum energy which allows elastic displacement of atoms from the site in NaCl lattice, amounts to 0.36 MeV (for neutrons 291 eV).

The accumulation of radiation defects in AHCs in the course of irradiation indicates that during division F and H centers, firstly come apart at the distance preventing their reverse reintegration and secondly both components of Frenkel pair are stabilized. If F centers do not require a special stabilization mechanism under room temperature, H centers are quite movable and their stabilization occurs on the account of complicated centers which include halogen atoms. Molecular ions $(X_3^{-})_{ac}$ (V$_2$ or V$_3$ center according to [97] or [98], respectively) are stable defects at room temperature in pure crystals. For example, Cl$_3^{-}$ centers are stable up to 90 °C in KCl [97]. According to data of Baimakhanov *et al.* [99] the stabilization mechanism of H centers in KCl depends on irradiation temperature. At temperatures below 200 K interstitial halogen atoms are stabilized with the formation of hole-type point defects (homogeneous defect formation) and at 300 K their major part drifts out to dislocations, dislocation loops, internal and external surfaces (heterogeneous defect formation).

The efficiency of Frenkel pair formation (the accumulation of F centers) depends on temperature. At reduced temperatures energy of Frenkel pair formation (E_{Fp}) for various AHCs can amount from several thousand to several hundred thousand electron volts. Under elevated temperatures (about room temperature) it approximates 1000 eV for all AHCs. The magnitude of E_{Fp} depends strongly on the irradiation power and is increased with dE/dx growth [100, 101].

The energy stored during radiation defect formation can be released in the result of the further annealing. Estimates carried out for AHCs demonstrate that around 10^{16} eV (0.1 % approximately) are released from 1 J of energy absorbed.

2.2.2. Electronic Color Centers

F center is the main primary electronic center in AHC. The method of electron-nuclear resonance demonstrated that the electron density in F center is distributed similar to electron density of hydrogen atom in ground state. The transition of electron from ground $1s$-state into the first excited $2p$-state conditions the absorption band of F centers. F center concentration can be defined by the Smakula formula:

$$N_F = 8.7 \cdot 10^{20} \frac{n}{(n^2 - 2)^2} \frac{k_m}{f} h \ (\text{m}^{-3}).$$

(2.7)

Where n is the refraction coefficient, k_m is the absorption coefficient in maximum of F band (m^{-1}), f is the oscillator strength, h is the half-width of the absorption F band (eV). Calculations indicate that during F center formation in NaCl lattice relaxation in the nearest two spheres amounts to $\Delta \approx$ -1-3 % [102]. In the third sphere $\Delta \approx 0.3$ %.

An important characteristic of F centers is their thermal stability which is depended on the activation energy of thermal destruction (E_s). Values of E_s and the destruction temperature of F centers for some AHC are represented in Table **2.3**. In most of the AHCs F centers are stable at room temperature and their accumulation is observed in the irradiation process. Generally accumulation kinetics is subdivided into two stages. In some cases - into three. Mathematical description of the accumulation curve is represented in the review [103]. At the first (fast) stage, F centers are formed by localization of thermalized electrons on the vacancies that existed before irradiation. The accumulation process transits into saturation under doses of several hundred grey. The convincing evidence of such mechanism of F center accumulation at the first stage is non-availability of the first stage in whisker crystals where vacancy concentration is very low [104]. The second stage of accumulation is conditioned by radiation disintegration of vacancy complexes with the further capture of electrons by released anionic vacancies. The third stage is connected to the Frenkel pair generation. F center accumulation curves in potassium halides irradiated by γ-radiation at room temperature are represented in Fig. (**2.5**) (our data).

Table 2.3. Thermal Destruction Energy of F Center (E_d, eV) and Destruction Temperature (T_d, K) for Some AHCs [105]

Crystal	LiF	NaCl	KCl	KBr	KI	CsBr
E_d, eV	0.16	0.074	0.15	0.135	0.11	0.04
T_d, K	700	480	570	500	400	430 [106]

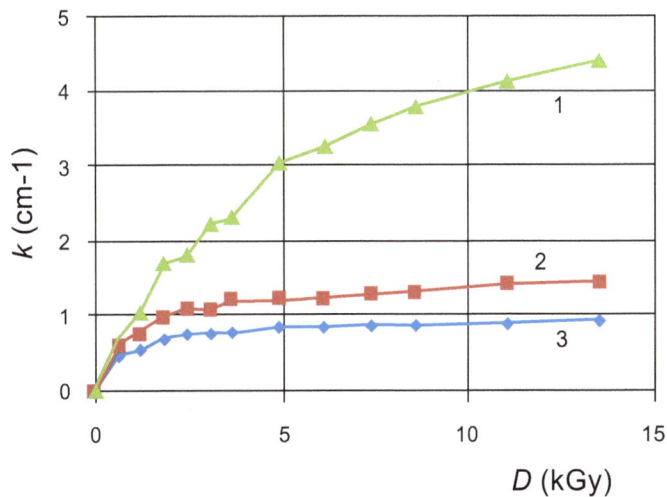

Fig. (2.5). The dependence of the optical density at the maximum F-band absorption at room temperature (γ-irradiation, $P_d = 0.9$ Gy s^{-1}). 1: KCl, 2: KBr and 3: KI.

In the connection with various mobility and temperature stability of primary radiation defects temperature dependence of F centers has a complicated appearance. Maximum effeciency of F center generation for KCl and KBr is achieved at ~ 200 K, and at 240 K for KI [107]. The temperature dependence of F center formation for potassium bromide is shown in Fig. (**2.6**) [108]. At low temperatures due to the small mobility of Frenkel pair components their separation is difficult. The efficiency of F center formation increases with rise in temperature that is connected with the increase of recession of F and H centers from each other. At elevated temperatures, the high migration rate provides an intense mutual recombination of the components of Frenkel pairs and their entry into the sinks.

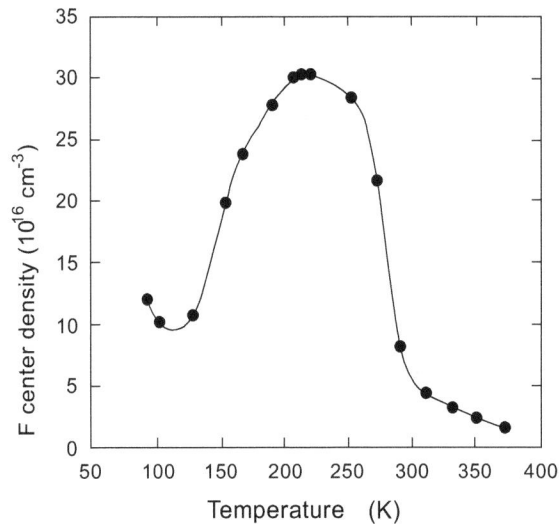

Fig. (2.6). Dependence of F center concentration on the temperature for KBr irradiated with excimer laser (4 10^5 MW m^{-2}, within two hours) [108].

Irradiation of alkali halide crystals at low temperatures can lead to the formation of F$^-$ center (sometimes it is called F' center) which is an anion vacancy that has captured two electrons. F$^-$center absorption band is located in long-wavelength spectrum area from F center absorption and has large half-width. The F$^-$ centers are destroyed at temperatures below room temperature (for example, in LiF at 160 K [109]). In additive and electrolytically colored AHC F$^-$ centers are well formed even at room temperature by highlighting in F-band.

In the case of long-time irradiation in long-wave spectrum area the additional absorption bands are observed apart from F-band which are conditioned by aggregate electron centers. Various size associates of F centers are included into aggregate electron centers: M center (double F center), R center (F$_3$ center), N center (F$_4$ center), X centers (pre-colloidal centers) and colloidal centers (atomic particles of alkali metal). M center as well as F center can capture one electron with the formation of M$^-$ center (M$^{'}$ center) [110]. Absorption spectrum of irradiated crystal KCl at room temperature with indication of some absorption bands is represented in Fig. (**2.7**). In the irradiated crystal without special treatment the concentration ratio of F, F$_2$ and F$_3$ centers corresponds approximately to the ratio n_F, n_F^2 and n_F^{3}, respectively. Photobleaching in absorption F-band, long-time irradiation at increased temperature and both of them result in dramatic reduction of single F center concentration and in the formation of large number of aggregate electronic centers.

2.2.3. Hole Color Centers

Hole centers include both the electron excitations and structural damage comprising one or several holes. Thermalized hole localized on two nearest halogen ions is the elementary hole center. Such electronic defect was called self-trapped hole or V$_k$ center. It was established the V$_k$ center is the two-atom molecular ion $(X_2)_{aa}$, occupying two anionic sites and oriented in the direction <110>. Meanwhile the distance between haloids is reduced by one third approximately. V$_k$ centers are very mobile at room temperature and are either captured by electronic defects or went out to dislocations, to internal or external surfaces. Their mobility drops with temperature decrease and they are "frozen" at quite low temperatures. Table **2.4** represents the decay (de-localization) temperatures of V_k centers.

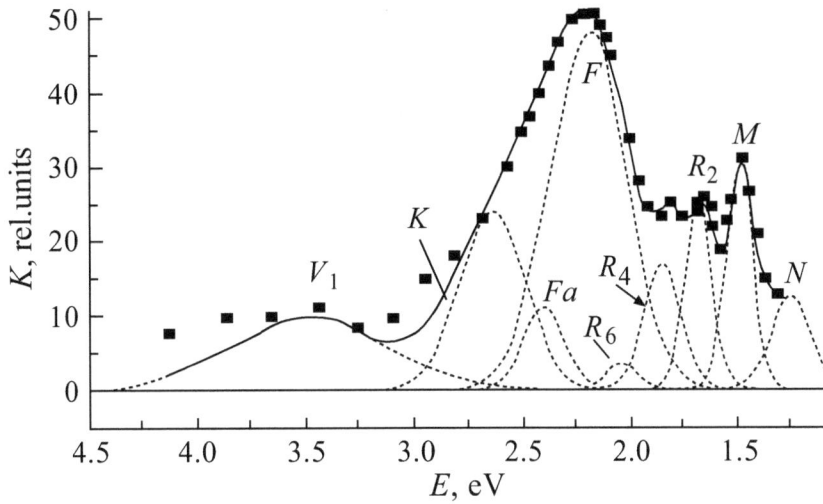

Fig. **(2.7).** Experimental (points) and calculated (solid line) absorption spectrum and the absorption bands (dashed lines) color centers for KCl irradiated with protons (100 keV, $8.6 \ 10^{18} \ m^{-2}$) [111].

Table 2.4: The Temperature Delocalization of V_k Centers (T_h, K) [112]

Crystal	LiF	NaCl	KCl	KBr	KI	RbCl	RbBr	RbI	CsCl	CsBr	CsI
T_h	125	150	210	160	105	220	170	125	195	145	90

Apart from V_k center hole centers also include V_F center (a hole localized in the cation vacancy), H center, V_2 center (three-halide molecule occupying two anionic and one cationic vacancy). The formation of double H centers was observed after X-ray irradiation in KBr even at 4.2 K [113]. The models of some hole centers are presented in Fig. **(2.8)** [114]. The molecular hole centers are enlarged in the course of long-time irradiation. The growth of molecular centers accounts for the sequential capture of H centers by V_2 center. Rzepka *et al.* [115] have shown the process of transforming in KI irradiated with X-rays.

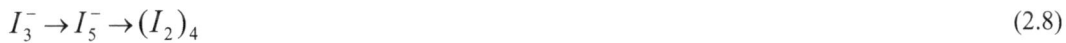

$$I_3^- \rightarrow I_5^- \rightarrow (I_2)_4 \qquad\qquad (2.8)$$

2.2.4. Defect Clusters and Radiolysis

The aggregation of radiation defects occurs under high-dose irradiation in AHC. If the formation of radiation defect clusters causes a change in the chemical composition of matter, then such a phenomenon is called ***radiolysis***. It was ascertained that in some cases intense irradiation results in voids formation without radiolysis. Thus according to Castro data [116] after γ-irradiation ($P_d = 15$ Gy s^{-1}, $D = 150$ MGy) isotropic cubical expansion $\Delta V/V = 3 \ 10^{-3}$ for NaCl and $4.5 \ 10^{-3}$ for LiF is observed. Swelling occurs mainly due to Shottky defect and their associate formation not due to the defects connected to F centers. According to Dubinko *et al.* [117] opinion when H centers are captured by spatial defects (mainly by dislocations) the emission of Shottky defects occurs according to the mechanism suggested by Seitz [96]. The energy emitted during exciton annihilation or during recombination of electron-hole pair on dislocation causes movement of dislocation jogs and Shottky defects are formed in the result of the same. Formation of intrinsic radiation defect aggregates of the size from tens to hundreds of nanometers in lithium fluoride was detected during irradiation by γ-rays by the method of diffuse X-ray scattering [118]. Vacancy associates in planes {200} are formed at doses of ~ 100 kGy. Aggregates of interstitial defects and their coalescence process shall be observed at doses of about 500 kGy. Inversion of vacancy associates into interstitial occurs in intermediate absorbed doses. Micro-void formation can occur in two stages. Thus the irradiation of CsBr at room temperature by γ-radiation or by powerful pulsed electron irradiation up to the dose of 5 10^7 Gy causes only a intense coloring of samples. The subsequent 10-minutes heating at 700 K leads to annealing of all color centers and to micro-void formation of the size 3 μm [119].

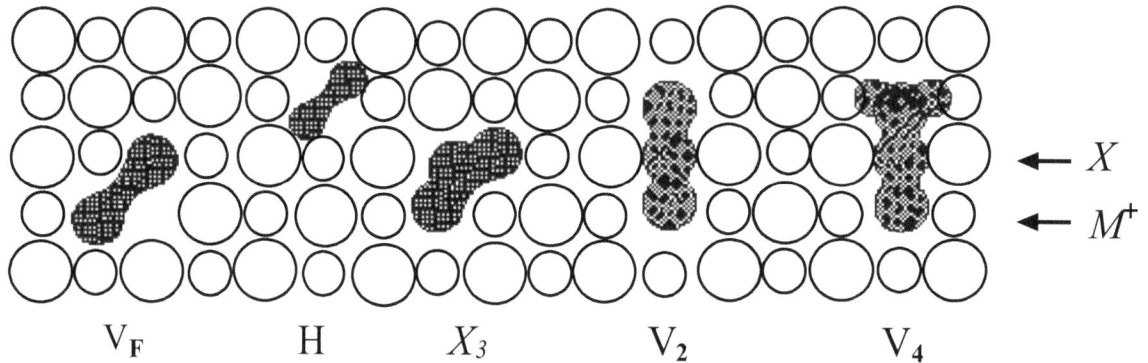

Fig. (2.8). Structures of hole color centers in AHC [114].

Not only vacancy cluster but halogen clusters as well are formed in AHCs during irradiation. Analysis of Raman scattering spectrum indicated that aggregates in the form of molecular ions X_n^- ($n > 3$), which size increases with the increase of the dose absorbed are formed in irradiated KBr and KI. Further aggregation of molecular halogen can result in dislocation loop formation [120]. Hobbs *et al.* [121] demonstrated that dislocation loops are formed during irradiation at the temperature when cationic vacancies are immovable. Formation of dislocation loops was observed by Baimakhanov *et al.* [99] after irradiation by X-rays in KCl at 200 K and Andronikashvili *et al.* [122] in LiF after irradiation by neutrons at room temperature by fluence of $\Phi = 10^{26}$ m^{-2} and further heating under the temperature of 450 °C. However, the dislocation loops are not formed if the absorbed dose does not exceed a certain value. Thus dislocation loops are not formed in NaCl under the γ-irradiation dose below 100 kGy ($P_d = 27.8$ Gy s^{-1}) [123].

Usually intense irradiation of AHC results in radiolysis. Numerous experimental investigations have established that radiolysis is possible only when equilibrium concentration of F centers exceeds some critical value (for example, approximately 3 10^{23} m^{-3} for NaCl). Moreover sufficient mobility of F centers or anionic vacancies shall be the required condition. The temperature of maximum efficiency of colloid center formation in NaCl shall be 150-170 °C (γ-rays, $P_d = 33.3$ Gy s^{-1}) [124]. At the same time the value of equilibrium concentration for F centers depends on the dose rate. The formation of atomic metal occurs in the transformation process of coherent to matrix X centers into non-coherent particles of alkali metals. At temperatures of 300-360 K the absorption dose required for X centers in NaCl equals to 1-5 MGy [125]. The efficiency of X center formation depends on the ratio of alkali metal atom radius to halogen ion radius (r_M/r_X^-). When r_M/r_X^- is decreased the X center stability is increased. For example, at room temperatures more X centers are formed in NaCl than in KCl under isodose irradiation. Impurities affect significantly on the formation of aggregate electron centers. So the divalent cations considerably increase the efficiency of F center formation, and inhibit the formation of X centers. Solute elements of some univalent cations (for example, silver in KCl-Ag) promote colloids formation. Impurities OH$^-$, O^{2-} and some other impurities in anionic sublattice also stimulate the colloid formation process.

The lithium fluoride is distinguished among AHCs for its high radiation stability. Irradiation by electron flows up to the dose 10^{20} m^{-2} (9 MeV, $\phi = 10^{17}$ m^{-2} s^{-1}) at room temperature results only in formation of isolated point defects [126]. Intense formation of radiation defects of the size up to 5-10 nm can be observed at the doses of 10^{21}-10^{22} m^{-2}. Annealing of such samples at 500-700 °C causes the growth of clusters.

The literature represents a number of design works simulating radiation defect formation from primary processes to the formation of alkali metal and molecular halogen (for example, [117, 127]). Dubinko *et al.* [117] calculated, on the basis of model of radiation-induced emission of Schottky defects, dependence of colloid formation efficiency in NaCl on irradiation temperature and dose absorbed. The data for the fixed calculation parameters (radius of colloid particle is 5 nm, concentration of colloid particles is 2 10^{22} m^{-3}, dislocations density is 5 10^{14} m^{-2}) are presented in

Fig. (**2.9**). The calculations showed that the growth rate can be positive or negative depending on irradiation conditions by ionizing radiation (not cascaded).

Interesting results were obtained by Schwartz and colleagues [91] under irradiation of LiF by lead ions with energy 5.9 MeV. Colloid formation could be observed even at 15 K, i.e. at the temperature when F centers are immovable. The formation of atomic metal occurred either on the account of F centers movement and aggregation within track spaces or in the result of direct capture of electrons by lithium ions with their further adhesion.

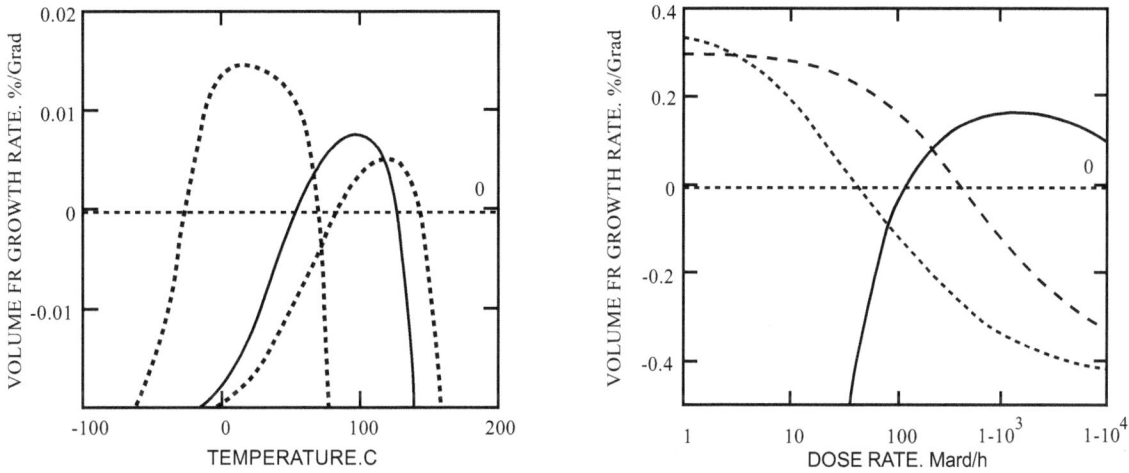

Fig. (2.9) Dependence of the growth rate of the colloid volume fraction (ΔV_c) during irradiation as a function of the temperature and the dose rate (P_d) for nominally pure NaCl: for fixed material parameters $r_c = 5$ nm, $n_c = 2 \ 10^{22}$ m^{-3}, $\rho_d = 5 \ 10^{14}$ m^{-2} (r_c and n_c is radius and colloid particle concentration, respectively, ρ_d is density of dislocations). 1: 2 kGy h^{-1}, 2: 2.4 MGy h^{-1}, 3: 10 MGy h^{-1}, 4: 0 °C, 5: room temperature, 6: 100 °C [117].

CHAPTER 3

Impurity Radiation Defects

Abstract: The models and thermal stability of the impurity radiation defects in the metals and alkali halides are reviewed. Due to the ionic lattice structure and presence of two sublattices in alkali halides, the formation of much larger number of types of impurity radiation defects than in metals is possible. The most important in the radiation processes are the complexes of self-interstitial-impurity and vacancy-impurity.

Keywords: Radiation, metal, alkali halide, impurity radiation defects, solute-point defects, defect complexes.

Impurities play an extremely important role in the processes of radiation-induced transformations in crystal solid solutions. Any impurity atom causes some distortion to the crystal lattice of the matrix. These distortions are the reason of the interaction of impurity atoms with radiation matrix defects. If in the result of such interaction stable complex is originated, then we can speak about the formation of impurity radiation defect. In general, the impurity radiation defect is any state of impurity, which occurs under irradiation and which differs from its initial state.

3.1. METALS

The most elemental impurity defects are complexes of solute-vacancy and solute-interstitial. Experimental data and theoretical evaluations demonstrate that formation of either complex mainly depends on the relative size of the solute atom. Naturally large size impurities tend to capture vacancies and smaller size impurities capture rather matrix interstitial atoms. An overwhelming number of the solute-point defect complexes in metals are stable at temperatures below room temperature.

3.1.1. Solute-Vacancy Complexes

Van der Kolk *et al.* [128] investigated formation and behavior of solute-vacancy complexes in tungsten. The complex formation occurred for oversized Ag, Cu, Mn, Cr, In implantation and during further annealing of implanted samples. It appeared that at temperatures when vacancies become movable, approximately a half of atoms implanted into tungsten capture one or more vacancies. Models of the supposed complexes are represented in Fig. (**3.1**). A binding energy of small complexes of solute-vacancy in tungsten amounts to 0.5-1.0 eV depending on the complex type. At temperatures exceeding annealing stage III, vacancy complexes are decayed. The most stable complex, supposedly, containing 4 vacancies has the binding energy 3.6 eV and exists at up to 1400 K. In metals with fcc structure, the solute atom and vacancy in the form of complex are located in the nearest lattice sites in the direction <110>.

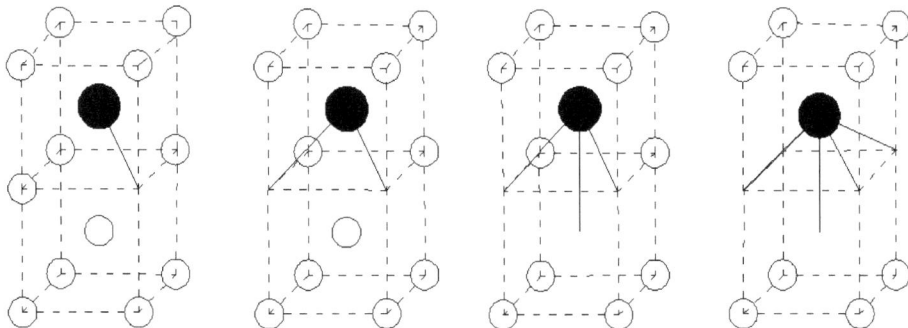

Fig. (3.1). Structures of solute-vacancy complexes [129].

Both theory and experimental data indicate that the binding energy of a simple solute-vacancy complex (E_{bV}) for metals lies in the range of 0.1-1.1 eV. Thus the binding energy of the complex iron-vacancy in aluminum is equal to 0.10-0.13 eV [129]. In nickel for the complex Y–vacancy and Ge-vacancy are $E_{bV} \approx 0.3$ and ≈ 0.17 eV and disintegration temperatures are 500 and 115 K, respectively [130]. According to Hacket *et al.* [131] calculations of the value of E_{bV} in alloy Fe-Cr-Ni for Zr, Hf, Ti and Pt are equal to 1.08, 0.71, 0.39 and 0.31 eV, respectively.

3.1.2. Solute-Interstitial Complexes

Proceeding from the ratio of the solute size to matrix atom size it should be suggested that the capture of self-interstitial (SI) by undersized solute is more probable than by the oversized solute. In this respect, Maury *et al.* [132] divide all alloys into two groups. In alloys of the first type (Cu-Au, for example) where solute atoms are of larger size than the matrix atoms, the SI do not form stable complex with the solute. The interaction of the solute with SI becomes apparent only under high solute concentration. Such interaction is expressed in SI mobility decrease. In alloys of the second type (Ag-Zn, for example), undersized solutes and SI form the complexes which are stable at lower temperatures. This is certified by experimental data of various authors. According to Dimitrov *et al.* [133] in Al-Cr alloy the solute-interstitial complex acts as combined dumbbell. In this case the chrome atom is slightly dislocated to the side of octahedral interstitial. Other undersized solutes (Cr, V and Ti) are also traps for aluminum interstitial atoms. Solute-interstitial complexes for Cr and V are destroyed at 200 K and at 110 K for Ti.

Swanson and Maury [77] have resulted another example. They established that during irradiation of aluminum with undersized solutes (Mn, Zn or Ag) at ~ 50 K, SIs are effectively captured by solutes with mixed dumbbell formation. Formation of mixed dumbbells with the participation of oversized solute (Sn) was not established. Structure of the solute-interstitial complex as the mixed dumbbell was proved by Mansel and Vogl [134] with the help of Mössbauer spectroscopy measurements of Al-Co alloy irradiated by neutrons. The complex formation occurs during irradiation at 4.6 K. Then the growth of the complex concentration is observed during heating of the irradiated sample at annealing stage I (~ 50 K). The complex is destroyed at temperatures above 160 K (annealing stage III) when the vacancies become mobile. It has also been established that the direct capture space of SI by cobalt atom amounts to 200 atomic volumes, i.e. the capture radius ~ 1 nm. It is also assumed that the complex Co-two SIs are formed.

With the help of the molecular dynamics method, Lam *et al.* [135, 136] completed calculation of structures and binding energies of the solute-SI (E_b) complexes in dilute alloys based on aluminum. The calculations demonstrated that solutes of smaller and larger size can form solute-SI complexes. The most stable structures for such complexes are the structures where the solute atom is located in equatorial plane to SI dumbbell in the first nearest site (*1b*) or in the second position (*2b*) (see Fig. (3.2)). Configuration (*2a*) is less stable. Octahedral position is the most stable for Be (E_b = 0.89 eV in the center of elemental cell with displacement by $a_0/10$ in the direction <100>); for Li, the most stable position is a tetrahedral interstice. The interaction between Mg atom and SI in Al-Mg alloy is very weak. Mixed dumbbell is the most stable interstitial state for Zn in Al (E_b = 0.38 eV) [137]. Meanwhile the calculations showed that migration energy of the mixed dumbbell equals to 1.2 eV which is much higher than 0.15 eV required for SI movement. On the other hand Abe and Kuramoto [138] established that mobility of the mixed interstitial dumbbell in dilute solutions of iron with Mo, Cr, Si, Be is higher than SI.

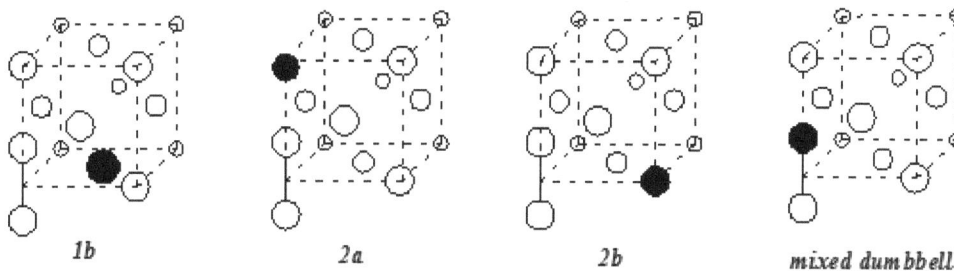

1b   *2a*   *2b*   *mixed dumbbell*

Fig. (3.2). Configurations of most stable interstitial-solute complexes in Al-based alloys.

Rehn *et al.* [139] informed about the existence of at least three various configurations of the complex Fe-SI in the alloy Al-Fe irradiated by electrons after annealing at 80 K. All three types of the complexes are very stable and are destroyed only at annealing stage III as a result of recombination with vacancies which become mobile at these temperatures. The binding and migration energies exceeded 0.63 eV for all complexes. The binding energies of complexes in aluminum and copper are represented in Table **3.1** and **3.2**, respectively.

Table 3.1. Binding Energy of Solute-Interstitial Complex for Alloys of Al (eV) [135, 136]

	Al-Cu	Al-Ag	Al-Au	Al-Ca	Al-K
1b	0.65	0.81	0.76	0.65	0.62
2b	0.44	0.74	0.87		
2a	0.41	0.53	0.50	0.47	0.51

According to Dworschak *et al.* data [140], effective radius of interstitial capture by any solute in aluminum irradiated by electrons (3 MeV, $\phi = 6.25 \ 10^{18} \ m^{-2} \ s^{-1}$) is reduced with the increase of irradiation temperature in the range 100-180 K. Magnesium loses its ability to capture interstitial atom in this temperature range, whereas other solutes (Ag, Ge, Zn) capture interstitial atoms even at 178 K.

The mixed dumbbells unite in some cases. Based on the investigation results of alloys Fe-(2-16)Cr irradiated by 5 MeV electrons, Nikolaev *et al.* [76] drew the conclusion on the existence of not only mono-mixed dumbbell with chrome participation (oversize solute) but of the double mixed dumbbell. According to Maury *et al.* [141], double mixed interstitial defects are formed in the Ag-Zn alloy too.

Table 3.2. Binding Energy of Solute-Interstitial Complex (E_b) for Alloys of Cu (eV) [142]

Element	Mg	Ti	Cr	Mn	Fe	Co	Ni	Zn	Ag	Au
E_b, eV	0.43	0.40	0.36	0.32	<0.20	<0.15	<0.15	<0.20	0.40	0.33

Maury *et al.* [143] presented the results of detailed investigation of oversized solute interaction with SIs in alloys based on iron. Several alloys, including Fe-Ti, Fe-V and Fe-Mo, with solute concentration within the range $5 \ 10^{-5}$-$3 \ 10^{-2}$ were irradiated by 1.6 MeV electrons ($\Phi = 7.4 \ 10^{21} \ m^{-2}$) at 25-35 K. The following conclusions are drawn upon the basis of the results obtained: (i) Iron SIs generated by electron irradiation are captured by all oversized solutes at the first annealing stage. (ii) The formed complexes are not mixed dumbbell and most likely are solute atoms, nearby SIs (see Fig. (**3.2**)). (iii) Interstitial iron atoms are released from complexes at annealing stage II at the temperatures depending on the solute size (see Table **3.3**).

Table 3.3. Dissociation Temperatures of Solute-Interstitials (T_d) and Radiuses of Some Oversize Solute in Iron (nm) [143]

Solute	V	Mo	Au	Ti
T_d, K	140	155	160	180
Pauling metallic radius, nm	0.134	0.139	0.144	0.147

Maury *et al.* [144] also studied the properties of interstitial atoms in dilute Fe-Ni, Fe-Mn and Fe-Cu alloys. The difference between the sizes of solutes in the specified alloys is insignificant. This is why they formed mixed interstitial defects during irradiation. In addition mixed interstitial defects are more movable than SIs and are quite stable. In the result of migration they form small mixed interstitial associates already at stage I. These associates are also mobile, so that the annealing stage II leads to the formation of small clusters. Such process causes the formation of solute precipitates, which are stable at room temperature in the Fe-Mn alloy with high solute concentration.

Arbuzov *et al.* [145] have established that in nickel, saturated with heavy hydrogen or tritium, the D-SI and T-SI complexes are formed after irradiation by 5 MeV electrons at 80 K. The binding energy of the complexes is 0.33 eV and they are destroyed at 150 K.

It should be noted that the addition of small amount of any impurities can significantly change the formation efficiency of the main solute-point defect complexes. For example, Swanson *et al.* [146] proved that addition of magnesium into Al-Ag alloy results in not only the decrease of mixed dumbbell (Ag-SI) formation efficiency within the temperature range 30-100 K but also in the change of annealing mechanism of interstitial defects at temperatures 100-240 K (annealing stages II and III). The reduction of efficiency of Ag-SI complex formation is connected to competition in SI capture between silver and magnesium at irradiation temperature up to 100 K. When temperature is increased at first, the release of SIs from Mg-SI complexes is implemented and then annihilation of Ag-SI mixed dumbbells with vacancies is completed. This process is preferable to implement the same on the account of movement of complexes and not of vacancies.

3.2. ALKALI HALIDE CRYSTALS

Unlike metals, a significantly larger number of impurity radiation defects are formed in AHC. The variety in the defect types is conditioned by the formation possibility of stable impurity states with variant valency and the presence of two sublattices. Radiation defect in AHC can be any impurity ion which has captured or lost one or several electrons as a result of irradiation. As a result of electron capture, the ***electron defects (electron color centers)*** are formed and the ***hole defects (hole color centers)*** are formed due to ionization or hole capture. The formation of a large number of stable radiation defect types along with the defect generation by under-threshold mechanism is the reason of low radiation resistance of AHCs.

3.2.1. Electronic and Hole Defects

Atomic ***E centers*** (Ag^0, Tl^0, In^0 and *etc.*) formed during electron capture by impurity ions M^+ [147], and ions M^+, formed during electron capture by divalent cation are included into the electron color center. An example of the latter case can be the Zn^+ ion formation in irradiated KCl-Zn crystals [97]. Some impurity ions can capture two electrons during ionizing irradiation or during highlighting in F absorption band after irradiation. ***B centers*** (Cu^-, Ag^- [148, 149]) and Pb^0 center are formed by this mechanism. Due to the charge sign change and in virtue of the fact that the size of negative metal ion exceeds the matrix cation size, such defects can be located not in cationic but in the anionic sublattice.

Similar phenomena are observed in the irradiated halide crystals with hydrogen impurity. Ion H^+ under ionizing irradiation can capture one or two electrons. After one electron capture hydrogen atom transits into interstitial (***U$_2$ center***) and after two electrons capture, it transits into anionic lattice with the formation of the so-called ***U center*** (H_a^-) [105, 150]. For example, ions H^- after irradiation occupy places of fluorine ions in crystals LiF and NaF [151]. U centers are effectively formed during irradiation in alkali halides doped with hydroxyl ions. The plastic deformation affects the stabilization form of single hydrogen centers. H_a^- centers are mainly formed in non-deformed LiF-OH crystals under irradiation but in deformed samples, the number of hydrogen anions and hydrogen atoms is the same. Meanwhile, hydrogen atom transition from anionic to cationic sublattice ($H_a^- \rightarrow H_c^-$) is possible [152].

In most cases, the type of impurity defect in AHC formed under irradiation depends more on its electron-acceptor properties than on the size. Metal cations of large electronegativity capture electrons effectively. Mg^{2+}, Pb^{2+}, Ni^{2+}, Mn^{2+}, Ag^+, Ga^+ *etc.* are included therein. For example Mg^+ and Mg^0 center formation was reliably established [153]. Bosi and Nimis [154] consider that the criterion of electron capture by divalence cation is the value of the second ionization potential. For the group of divalent cations mentioned above $\varepsilon_2 > 15$ eV. Alkali, alkali-earth and rare-earth metals have smaller electronegativity ($\varepsilon_2 < 15$ eV) and electrons are not captured at room temperature. Ion Bi^{3+} does not capture electrons in irradiated NaCl-Bi crystal both at room temperature and at 77 K [155].

It is asserted in most works devoted to AHC doped by europium that Eu^{2+} ion does not capture electron. For example, according to Nierzewski *et al.* data [156] europium is in the single state of Eu^{2+} in crystals KCl-Eu under irradiation within the temperature range 80-300 K. At the same time in [157, 158] the conclusion was drawn that Eu^+ ions are formed in KI-Eu.

Increased affinity to electron of Ni^{2+} causes, for example, not F_H center formation but electron transition from F center to nickel ion with Ni^+ center formation [159]. Nickel electron-acceptor properties are so vivid that it easily transits into atomic state under irradiation and forms a metal colloid.

Gold, which has a high value of electron affinity (2.3 eV), forms a stable Au^- ion located in the anionic sublattice of the CsBr crystal. Irradiation of CsBr-Au crystals by ultra-violet light results in the formation of gold ion aggregates. Large aggregates of Au^- ions have absorption under 579 nm and are similar to X centers in AHCs [160].

Hole defects appear as a result of direct ionization or capture of a hole. Existence of Tl^{2+}, In^{2+}, Cu^{2+} centers formed from initial ions M^+ is well established. These centers are stable up to temperatures slightly below room temperature. For example, in NaI-Tl, the ion Tl^{2+} is destroyed at 236 K [161]. Ions Pb^{2+} and Sn^{2+} can increase their valency per unit under irradiation. There are contradictive data on the possibility of stable Eu^{3+} ion formation in AHCs. The valency increase in alkali-earth ions has not been established.

Formation of mixed hole centers was established in anionic sub-lattice of AHC doped by anionic homologues. Irradiation of such crystals causes the formation of the mixed self-trapped holes (mixed-halogen V_k centers). For example, Goldberg [162] observed the emission of mixed ICl^- molecule under X-ray irradiation of KCl-I crystal.

3.2.2. Impurity-Radiation Defect Complexes

If the defects of the crystal lattice caused by impurity and radiation defect compensate each other, they can be combined at collisions and form the complex impurity-radiation defect (IRD). In some cases of the lattice distortion caused by impurity ion is such as the formation of the primary radiation defect whereby the solute is energetically more efficient than in regular crystal sites. In this case IRD can arise directly in the moment of Frenkel pair formation. The well-studied IRD include the following:

(i) Defects which Include Interstitial Atom of Halogen

H_A center: H center nearby the alkali metal impurity ion. The binding energy of H_A center is small and, for example, for Na^+ in KBr amounts to 0.025 eV [163].

H_B center: H center nearby the oversized impurity halogen. H_B center is thermally more stable as well as H_A center as compared to H center [164].

H_z center: complex M^{2+}-H center. According to Baimakhanov *et al.* data [99] H_z center consists of M^{2+} ion and molecular ion $(X_2^-)_{ac.}$.

V_4 center: A complex including H center and impurity cation. If the impurity ion is homological cation then such defect is designated as V_{4A} center.

(ii) Defects which Include F Centers

F_A center: F center nearby the alkali metal impurity ion.

Z center or F_Z center: F center nearby the impurity M^{2+} ion.

F_H center: F center nearby the impurity anion. For example, $F_H(O^-)$ centers are formed in potassium halides with oxygen impurity. The O^- ion occupies non-central position of anion site of lattice [165].

F_d center: F center nearby the O^{2-}-V_a dipole.

F_A centers are subdivided into $F_A(I)$ and $F_A(II)$ centers depending on dimensional orientation of F center and alkali impurity ion [166]. F center-impurity complex is more stable thermally with a monovalent cation of non-alkali metal. For example, in KCl-Ag, such center is effectively formed at room temperature [167]. Meanwhile, due to high electron-acceptor properties of Ag^+, the electron is located on anionic vacancy only by 30-40 % and by 60-70 % on silver ion. In this connection, Ahlers *et al.* [167] considered that it is more correct to designate the center not as $F_A(Ag^+)$ center but as $\mathbf{A_F(Ag)}$ **center**.

The formation of complexes Cu^0-V_a^+ was established in KCl-Cu crystal under irradiation by X-rays. Meanwhile such defect formation occurs on the account of anionic vacancy migration to the nearest copper atom at temperature exceeding 220 K [168]. Complex Cu^0-V_a^+ can be formed in other ways: When electron is captured by Cu^+-V_a^+ associate or in recombination of V_F center with Cu_a^- anion.

Divalent solute cations can form both electron and hole complexes. Usually alkali-earth and rare-earth elements whose electron affinity is fairly small, do not capture electron themselves but rather form electron centers in the form of Z centers. Five types of Z center are established till now however there are variances in the interpretation of individual Z centers. Detailed review of Z center models is represented by Radhakrishna and Chowdari [169]. Only model of Z_1 center (complex including F center and dipole M^{2+}-V_c^-) can be considered as the generally accepted model. Z_1 center is formed in the time of irradiation in parallel with F centers in the majority of AHC-M^{2+} crystals at room temperature. Other types of Z centers are formed in irradiated samples after additional thermal treatment and/or after photo-highlighting. Thus for example Z_2 centers are formed in crystals LiF-Mg after heating up to 100 °C, and Z_3 centers are formed after heating at 125-175 °C by reaction $Z_2 \rightarrow Z_3 + e^-$ [170]. The temperature of thermal decay of Z centers falls within the temperature range 65-150 °C and is represented in Table **3.5**.

Bosi and Nimis [154] analyzed the possibility of Z_1 centers formation in alkali halides from classical physics positions. In their opinion, the criterion for Z_1 center formation may be the correlation between the ground energy state of impurity ion (E_a) and energy of F center ground state (E_F). The E_a value can be evaluated by the equation:

$$E_a = (h\nu)_+ + \frac{e^2}{d} - \varepsilon_2 + W,$$ (3.1)

where $(h\nu)_+$ is the energy of optical transition in the ion M^+, ε_2 is the second ionization potential, and W is the work of ionization halogen. If $E_a < E_F$, the reduction of M^{2+} ion to M^+ is preferable. If $E_a > E_F$, the formation of Z_1 center is expected. By using the suggested criterion it is possible to suppose that Z_1 centers can be formed in KCl under $\varepsilon_2 < 11.6$ eV, in KBr under $\varepsilon_2 < 11.3$ eV and in KI under $\varepsilon_2 < 10.7$ eV. The approach used permits to understand the experimental data when Z_1 centers are well formed in KCl. They are poorly formed in KBr and they are not formed in KI.

Irradiation in temperature range 80-200 K in crystals KCl-Sr mainly results in the formation of H_z centers [99]. These centers are destroyed at 250 K. Then a part of them captures one H center and transforms to $\mathbf{V_z}$ **centers**. Irradiation of KI-Pb crystals also leads to the H_z center formation [171]. The luminescence of Pb^{2+} ion perturbed by molecular ion $(I_3^-)_{aca}$ was registered in the same sample. Such complex was called $\mathbf{V_{2z}}$ **center**. Egemberdiev *et al.* [172] noted that in irradiated AHC with lead impurity, the formation of three types of impurity-defect complexes is possible. Each of the types includes anionic and cationic vacancy and the lead with various oxidation degrees (Pb^{2+}, Pb^+ and Pb^0). Several types of Mn^0_a centers (in anionic vacancy) were established for irradiated NaCl-Mn crystal [173]. Several types of solute complexes are formed under radiation in KCl-In crystal. In the result of ionizing irradiation from In^{3+}-Vc^- dipole, the H_A centers (In^+-$(Cl_2)_a^-$) and H_z-centers (In^{2+}-V_c^--$(Cl_2)_a^-$) as well as hole centers V_{2A} ($In^+Cl_3^-$) and V_{2z} ($In^{2+}Cl_3^-$) centres are formed [174].

Van Puymbroeck and Schoemaker [175] investigated though the method of electron-spin-resonance the capture of halogen interstitial atoms in KCl doped by divalent cations (Ca^{2+}, Ba^{2+}, Sr^{2+}, Cd^{2+}, Pb^{2+}, Mg^2, Sn^{2+}). It was shown that interstitial halogen atoms formed under affection of X-ray can be captured by both single M^{2+}-V_c^- dipoles and by their dimmers. Such impurity radiation defects are called H_D, not H_z centers by authors of [175].

The formation of H_{DD} **center** was detected in irradiated KBr-Sr crystal by the method of double electron-nuclear resonance [176]. H_{DD}-center is a complex which includes two strontium ions located in the nearest cationic sites along the axis <010> and self-trapped hole located along the axis <100> between them. The center is stable at room temperature.

Irradiation by X-rays of NaCl-Fe crystals leads to the complex formation with participation of Fe^{3+} ions [177]. Two models of such complex were suggested. At temperature below 200 K, the complex consists of Fe^{3+} ion replacing Na^+ ion, of cationic vacancy and of H center in the nearest neighboring position. At room temperature the ion transits into interstitial position and forms a complex with cationic vacancy and bivacancy (V_cV_a) along <100> axis. It should be noted that both complexes suggested are charged positively in relation to the crystal lattice.

The formation of $(F_2^+)_H$ centers was established in AHC with oxygen impurity. It is assumed that this center is the F_2^+ center (ionized M center) nearby O^{2-} or O^- ions [178, 179].

An interesting peculiarity was detected in γ-irradiated AHC doped with OH^- ions. It happens that irradiation at room temperature results in the formation of hydrogen bond between hydroxyl group and interstitial halogen atom ($OH\cdots Hal_{int}^0$) [180]. The formation of complexes $O^{2-}H^+\cdots F_2^-$ (V_{1OH} center) and $M^{2+}O^{2-}H^+\cdots F_2^-V_c^-$ (V_{1MOH} center) is also assumed [181].

The analysis of data available in literature testifies to the fact that mixed impurity complexes are formed in AHC doped with several solutes in the result of ionizing irradiation. For example, $H_i^0Mg^{2+}$ complexes (interstitial hydrogen atom nearby magnesium ion) were detected in irradiated LiF-H-Mg crystal [182].

3.2.3. Thermal Stability of Radiation Defects

Thermal stability is the important characteristic of impurity radiation defects. To provide for effective participation of radiation defect in radiation transformations, its lifetime at irradiation temperature shall be long enough. The larger the binding energy of the complex, the higher the temperature of the destruction of this defect (T_d). It can be considered that the defect is stable at $T < T_d$ and unstable at $T > T_d$. Data on E and Z center destruction temperature are summarized in Tables **3.4** and **3.5**.

Table 3.4. The Temperature of the Destruction of E Centers for Some Crystal Phosphors According to [177, 183, 184, 185-187] (K)

	Tl^0	In^0	Ag^0	Cu^0
NaCl	215, 220	185	240, 245	188, 225
NaBr	113		160	
KCl	300	240	370, 405	243, 261
KBr	185, 195	115	265	170
KI	180, 185	140, 047		240
RbCl	185, 300		380, 390	
RbI	185	145		
CsI	118	< 77		

Table 3.5. Destruction Temperature of Z Center for Some AHC-M^{2+} [189-199] ($^{\circ}$C)

System	NaF-Ca	NaCl-Ca	NaCl-Sm	NaCl-Sr	NaCl-Tb	KCl-Ca	KCl-Ba	KBr-Ca	RbBr-Ba
Z_1 center	114 [195]	107 [198]	130 [190]	120 [194]		65 [199]	75 [191]	65 [197]	65 [196]
Z_2 center			80 [189]		110 [192]	135 [193]			

It is clear from the data represented in Table **3.4** that in AHCs with the same anion, the temperature of E center destruction is higher when the difference in ionization potential of the matrix and impurity cations is greater. In AHCs with same cation, the E center stability decreases with the increase of ionicity degree of crystal lattice. It should be noted that atomic centers are stable at room temperature only in KCl with impurities of silver and thallium and in RbCl with silver impurity. It is interesting to note that if Tl^0 forms the complex with anionic vacancy in NaCl, it becomes stable at 370 K [188].

M^{2+} ions originating from monovalent cations are destroyed within the temperature range 200-400 K. For example, in NaI-Tl, the Tl^{2+} ion disappears at 236 K, and in NaCl-Cu Cu^{2+} at 391K [185].

The data in Table **3.5** suggest that the Z center is stable in all submitted AHCs at room temperature. Electronic defects, which contain electron-acceptor divalent cations, are also stable at room temperature. For example, Pb^+ centers are disintegrated in NaCl-Pb at 110 °C [200].

The temperature of the destruction of M^{3+} ions (ionized impurity M^{2+} cations) is near room temperature. For example, Pb^{3+} ion is disintegrated in KCl at 60 °C [201].

Complexes with the participation of divalent cations are more stable than F_A and H_A centers. Thermal stability of F_A centers depends on impurity size. The smaller ion size, the more stable F_A centers.

CHAPTER 4

Effect of Impurities on the Radiation Defect Formation

Abstract: The effect of impurity atoms on the efficiency of formation and accumulation of radiation defects in the metals and alkali halide crystals is considered. Interaction of impurities with radiation defects can both promote stabilization of the structural disorder, and suppress the radiation-defect formation. Especially the complex processes take place in the irradiated alkali halide crystals, where a wide variety of the radiation-induced defects is possible.

Keywords: Radiation, metal, alkali halide, impurities in crystals, formation efficiency of radiation defects, radiation resistance.

Influence of impurities on the radiation processes in solid solutions are connected with the following main phenomena: the effect of impurities on the formation of intrinsic radiation defects, the formation of impurity defects and the effect of impurity defects on the structure of solid solutions. The second case has been considered in Chapter 3, while the third one will be discussed in subsequent chapters. This chapter presents experimental data on the effect of impurities on the formation and accumulation of radiation defects.

4.1. METALS

Numerous experimental data show that the addition of any element in metals and alloys can either increase or decrease their radiation resistance. The magnitude and sign of the effect depend on the nature of solute, its concentration and the irradiation condition. Here are some examples of the effect of impurities on the void formation and swelling of steels.

Swelling of austenitic steels decreases drastically with excess concentration of nickel about 20 wt. %. On the contrary, a small concentration of phosphorus stimulates swelling. Or, a small amount of molybdenum promotes swelling, but its large quantity can cause an opposite effect. Dimitrov, C. and Dimitrov, O. [202] studied austenitic Fe-Cr-Ni alloys by electrical resistivity measurements during 21 K electron irradiation. It is found that increasing of the nickel concentration leads to a significant increase in the mobility of self-interstitials and a small decrease in the vacancy mobility.

It should be noted that not always the impurities with approximately same size (oversized or undersized) can affect the radiation processes in the alloys equally. For example, Nakata *et al.* [203] found that the addition of 0.3 wt. % of oversized titanium or zirconium in stainless steel 316 markedly reduces the void density after irradiation with 1 MeV electrons ($\phi = 3 \ 10^{23}$ m^{-2} s^{-1}) at 823 K to dose of 50 dpa, compared with that in standard 316 steel. In contrast, the addition of 0.3 wt. % of oversized vanadium increases the void formation under the same conditions of irradiation.

In some cases, the addition of a third element can significantly affect the behavior of the main solute under irradiation. For instance, Kato *et al.* [204] showed that the addition of zirconium decreased the grain boundary chromium depletion and nickel enrichment.

Mayer and Morris [205] presented data on the effect of irradiation by fast neutrons on the dilute solutions of aluminum with silicon or indium at pile temperature. It is found that indium enhances and silicon suppresses the nucleation of voids. This has explained the fact that indium is a trap of vacancies, and silicon precipitates are traps of both vacancies and interstitials.

4.1.1. Decrease of Radiation Resistance

The presence of iron 1.5 wt. % in copper distinctly affects the formation of voids under the irradiation with 650 keV electrons ($\phi = 3 \ 10^{23}$ m^{-2} s^{-1}) at 523 K [206]. Void formation has already started at doses of 1.5-2 dpa. Voids were formed preferentially near the coherent precipitation, and increased in size as the dose increased. Quenching substantially reduces void formation, but during irradiation, the radiation-induced precipitates stimulate the void formation.

Solutes which effectively form a solute-defect complex, in most cases reduce the radiation resistance of metals. For example, Ti, Sn, Dy and Au increase the defect creating in zirconium under neutron irradiation ($\phi = 1.3 \ 10^{16} \ m^{-2} \ s^{-1}$, $\Phi = 3.7 \ 10^{21} \ m^{-2}$) at 10 K [207].

Hydrogen plays a special role in the evolution of structural defects during irradiation. The presence of hydrogen lowers the radiation resistance of nickel to the swelling. It has been established [208] that when hydrogen atoms combine with vacancies, they become centers of the association of vacancies. Under intensive irradiation by hydrogen or helium ions the gas bubble sublattice can be formed [209]. Parameters of gas bubble sublattice do not depend on irradiation temperature. The approximate radius of the bubbles is of 1 nm, the distance between the bubbles is 3-9 nm.

4.1.2. Increase of Radiation Resistance

Studies by various authors show that the increase of carbon concentration in the alloys leads to a decrease in their swelling under irradiation. Leinaker *et al.* [210] indicate that carbon (and nitrogen), even in small amounts reduces the swelling of austenitic stainless steels and nickel at temperature below 550 °C, while at higher temperatures, the irradiation swelling may increase. The suppression of swelling can be explained by the fact that carbon capture vacancies. Habtetsion *et al.* [78] note that carbon affects the radiation processes in cobalt. It is found that if nickel and iron have little effect on the recovery curves of the resistivity for cobalt irradiated with 3 MeV electrons, the carbon strongly suppresses stages I and II of annealing. This indicates that carbon affects the shape of the stabilization of interstitial atoms.

Silicon (analog of carbon) plays an important role in radiation-induced transformations in the alloys. Averback and Ehrhart [211] found that silicon suppresses the growth of interstitial clusters under annealing of electron irradiated dilute alloys of nickel to 200 K. A large number of interstitial clusters, their low mobility and relatively high mobility of vacancies promote the suppression of swelling in Ni-Si alloy.

Titanium and rare earth elements are the impurities, which in most cases increase the radiation resistance of alloys. Long neutron irradiation (during the year at 480 °C) of austenitic steel with the addition of titanium does not lead to noticeable structural changes [212]. Wiffen and Maziasz [213] also noted that the introduction of titanium in the austenitic steel dramatically suppresses radiation swelling. Irradiation of mentioned steel by fast neutrons at 55 °C to a dose of 10.5 dpa does not cause swelling, voids and precipitates a new phase, but increases the dislocation density from 10^{10}-10^{12} to $1.5 \ 10^{15} \ m^{-2}$.

Platov *et al.* [214] found that the addition of scandium markedly increases the radiation resistance of aluminum-magnesium alloy under certain condition. The fast neutron irradiation of Al-2.5%Mg-0.4%Sc alloy no affect the tensile strength and yield strength in the temperature range of 20-300 °C ($\Phi = 1.3 \ 10^{24} \ m^{-2}$). At the same time the 2.3 MeV electron irradiation ($\Phi = 1.3 \ 10^{22} \ m^{-2}$) induces decomposition of solid solution at room temperature in the samples annealed at 350 °C with the formation of phase precipitates Mg_2Al_3, Mg_3Al_4 and Mg_5Al_8 [215].

The presence of Fe_3Al nano-particles (with the size of 4.5 nm) leads to a marked decrease of accumulation of vacancy clusters under 5 MeV electron irradiation ($\Phi = 5 \ 10^{22} \ m^{-2}$) of the alloy Fe-Ni-Al at temperature 300-573 K [216]. The presence of elastic stresses near the interface may be the one of reasons that lowers the number of vacancy clusters during irradiation, in spite of the coherence of the precipitates. These stresses cause the flow of vacancies to the precipitates. Another possible reason may be that formation energy of Frenkel pairs is lower in the nano-particle Fe_3Al than in the matrix. Unfortunately, it is extremely difficult to assess.

Arbuzov *et al.* [217] have established that the small additive of sulfur (0.1 %) suppresses the formation of vacancies in austenitic steel Fe-36%Ni during 5 MeV electrons irradiation at temperature 573 K.

4.2. ALKALI HALIDE CRYSTALS

4.2.1. Homological Impurity Effect

Alkali halides doped by undersize homologous cations and oversize homologous anions were mainly the object under research of radiation defect formation of AHCs with homologous impurities. Irradiation of AHCs doped by homology cations of smaller radius leads to the formation of F_A and V_{4A} centers. The efficiency of formation of intrinsic and extrinsic defects depends on the irradiation temperature and impurity concentration. At low temperatures, when the H centers are fixed, the impurity ions capture the excitons and suppress the formation of Frenkel pairs. According to Bekeshev *et al.* [218] the color center concentration is much lower in irradiated KBr-Li than in pure or weakly doped KBr at temperature 4.2 K. In temperature range of 50-230 K, the Li^+ ions capture the mobile H centers (forming the H_A centers) and thereby raising the efficiency of formation of defects. The generation efficiency of F_A, V_4 and V_{4A} centres increases sharply at temperatures above 230 K under increase of concentration of lithium to 10^{-4}. Besides, V_4 and V_3 centers are formed in pure or low doped KBr crystals, but the multi-halide V_3 centers are not formed in high doped KBr-Li during irradiation at temperature above 50 K. *I. e.* the lithium ions counteract the association of halogen atoms. This results from the fact that lithium, having a small size, occupies a noncentral position in cation site and is a trap for interstitial halogen. For the same reason, the origin of Frenkel pair close to a lithium ion is more favorable than in regular lattice site.

Sodium (other ion of undersize alkali metal) affects the radiation defect formation in KBr as well as lithium. According to Giuliani [219], sodium also raises the formation of the F centres and anion vacancies under X-ray irradiation in KBr at 80 K.

Still and Pooley [220] have established that at helium temperatures in cation mixed AHCs the efficiency of F centre formation is more low, than in pure AHCs. According to their data in KCl-RbCl under irradiation with 400 keV electrons ($\bullet = 1.4 \ 10^{16} \ m^{-2} \ s^{-1}$, $T = 5$ K), formation energy per one F centre increases from about 2200 eV in KCl and 18500 eV in RbCl to 43250 eV in $K_{0.16}Rb_{0.84}$-Cl. A significant reduction in F centre production rates can be expected in mixed alkali halide because of the disruption of replacement collision sequences.

Introduction of oversize homologous anion leads independently on temperature to the formation decrease of the color centers. For example, formation energy per one F centre increases with 2200 eV in KCl and 3000 eV in KBr to 10000 eV in $KCl_{0.4}$-$KBr_{0.6}$ under irradiation with 400 keV electrons at 5 K [221]. Besides, the oversize anions influence on types of the hole centers formed by ionizing radiation.

According to Korepanov *et al.* [164] in crystals KBr-I, the X-ray irradiation leads predominantly to the formation of the V_2 and V_4 centers, while in pure KBr the V_3 centres are mainly formed. That is, the oversize homologous anion suppresses association of the H centers just as in a case with undersize homologous cation. Hirai [221], Still and Pooley [220] suggested that the low efficiency of radiation colouring of AHC doped with an oversize homologous anion is connected with the difficulties of division into components of Frenkel pair when H (or I) center collides with anions of larger mass.

Homologous undersize cations and oversize anions can capture not only the interstitial defects and F-centers, but also hole and exciton. According to Nagli and Karklinja [222] the electronic excitations are localized in CsI-Na mainly near the activator, and Malyshev and Yakovlev [223] found that every third hole-electron pair is localized on the iodine ions at 80 K in KCl-I (0.17 mol % I).

4.2.2. Effect of Non-Homological Cations

Sufficient experimental data is now available to show that the doping of AHC by impurity cations leads to an increase in radiation-defect formation. For example, the efficiency of formation of F centers was higher in doped NaCl and KCl regardless of the type of cation impurities (Ag, Tl, Sr, Mn, Ca, Cd, Co) than in pure samples under electron irradiation at 80 K [224, 225]. However, the effect on the radiation colouring can differ greatly depending on the nature of the impurity. The strongest increase in F center concentration was observed in the initial stages of coloration at room temperature with the introduction of alkaline earth and rare earth elements in alkali halides.

Growth of the cation impurity concentration causes an increase of the F center formation rate. However, thermal and optical stability of F centers is lower in doped crystals than in pure.

A large number of works is devoted to the formation of radiation defects in AHC doped with europium. According to Shuraleva and Ivakhnenko [107], introduction of europium leads to a more effective F center formation in KCl in both the first and second stages of accumulation, as well as to shift the maximum efficiency of coloring in the region of high temperatures. The increase of F center accumulation rate is explained by the fact that Eu^{2+}-V_c dipoles capture halogen atoms. In this case one dipole can stabilize up to 10 H centers. Fig. (**4.1**) presents the curves of the F centre accumulation for KCl-Eu with different impurity concentration under γ-irradiation (our data).

According to Opyrchal *et al.* [226], efficiency of KCl-Eu coloring increases and tends to saturate with increasing concentration of europium under γ-irradiation ($P_d = 2.8$ Gy^{-1} s^{-1}, $D = 2.5$ kGy). The time required to reach saturation, is reduced with increasing of Eu concentration (n_{Eu}). It was found that the F center concentration at saturation is proportional to $(n_{Eu})^{0.33}$. From the received results it is concluded that coloring of KCl-Eu proceeds just as in potassium chloride doped with alkaline metals. The rise in the F center formation efficiency occurs because europium stabilizes H centers. However, calculations showed that H centers are captured by small impurity aggregates, and not single dipoles Eu^{2+}-V_c^-.

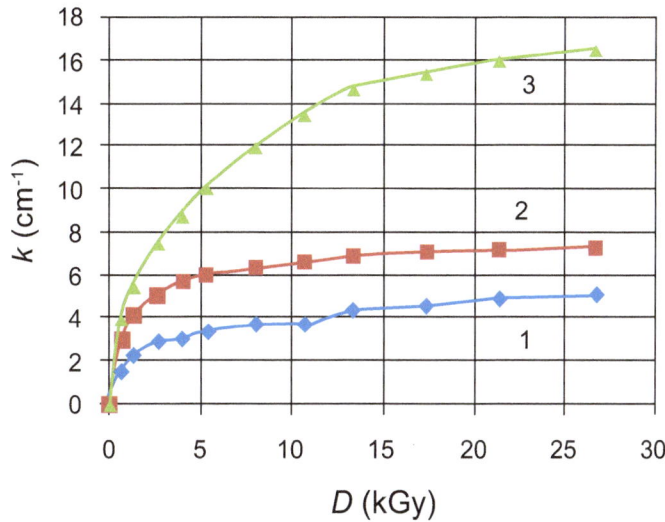

Fig. (4.1). Accumulation F center curves. 1: KCl; 2: KCl-Eu ($k_{Eu} = 1.8$ cm^{-1}) and 3: KCl-Eu ($k_{Eu} = 8.4$ cm^{-1}).

In a number of works [227-232] it is established that the efficiency of F center accumulation at room temperature is proportional to the square root of impurity concentration in AHCs doped with divalent cations (Eu^{2+}, Pb^{2+}, Ca^{2+}, Cd^{2+}) in the first stage of coloration. At the same time in potassium halides doped with Eu, the F center concentration is linear from absorbed dose with the onset of irradiation at 93 K, *i.e.* the first stage of accumulation is absent [157].

It should be noted that the radiation processes proceed in AHC doped with calcium a little differently, than in AHC doped with other alkaline earth metals. Investigation of NaCl-Ca colouring under γ-irradiation (about 400 R s^{-1}) showed [233] that the efficiency of F center formation in the first stage of coloring is not correlated with both impurity dipole and free vacancy concentration. The aggregative state of impurities also does not affect efficiency of radiation colouring of NaCl-Ca. However, a slight increase in the production rate obviously is connected to the first stage of dipole aggregation.

Rzepka *et al.* [115] studied the formation of V-type defects in KI and KI-Ca under X-ray irradiation. If V_2 centers are formed in the pure KI only on the second stage of coloration, then these centers arise already in the crystals of KI-Ca in the first stage. This difference is explained by the fact that cation vacancies formed in pure KI during irradiation, while in KI-Ca they have already existed before irradiation as Ca^{2+}-V_c^- dipoles. Prolonged irradiation of

KI crystals and KI-Ca causes to form poly-iodides. It is interesting to note that in KI-Tl the poly-iodides were not found under similar conditions of irradiation.

A more complex pattern in the radiation process is observed in alkali halides doped with electron-acceptor impurity (Ni, Pb, Tl, *etc.*). Dependence of the F center concentration on the content of such cation has maximum. According to [234, 235] the maximum is observed at a concentration corresponding to the solubility limit of these impurities. The high electron-acceptor ability of Ni^{2+} ions is the cause of decrease in the F center formation efficiency at both the first and second stages of accumulation. In the first case it is because of the competition for the capture of electrons from the anion vacancies, and the second one is due to the competition for the capture of H centers [159].

The results of studies of F center accumulation in alkali halides doped with lead are presented in [234-236]. It was found that for quenched samples the dependence of the F center number (N_F) on the lead concentration (n_{Pb}) passes through a maximum under room temperature irradiation, and for the aged samples N_F increases monotonically, reaches saturation (see Fig. (**4.2**)). The maximum of N_F is reached at n_{Pb} value which is the solubility of lead in the AHC. Increasing of F center accumulation in the beginning of irradiation is connected with stabilization of H centers by single Pb^{2+}-V_c^- dipoles and their aggregates. With increasing concentration of impurities in quenched samples, the number of complexes of impurity-H center increases, and the average distance between them and the F centers is reduced. This process leads to an increase of probability of mutual recombination of F and H centers, and consequently, to lowering of the F center concentration with increase of lead concentration. In aged crystals, the concentration of single Pb^{2+}-V_c^- dipoles does not depend almost on the n_{Pb} values and may even decrease with the increase in n_{Pb} values because of coalescence process. This is reflected in the dependence of $N_F = f(n_{Pb})$ in the second stage of colouring in the form of a plateau at high values n_{Pb} and as the independence of N_F on n_{Pb} in the first stage of coloring. Lowering the F center formation in the first stage of coloration in quenched crystals is connected with a competition for capture of electrons between the lead ions and pre-radiation anion vacancies. Similar results were obtained for quenched samples of NaCl-Mn irradiated with X-rays (the second stage of coloration) [237] and for γ-irradiated (104 Gy) KCl-Tl [238].

There are other assumptions about the mechanism of F centers concentration increase at the lead introduction in KCl. According to Gavrilov *at al.* [239] increase of the radiation colouring of KCl-Pb, as well as of KCl-Tl is not connected with the capture of H center by impurity ion, but with a predominant decay of exciton localized near the impurity ion, and with the subsequent stabilization of F center around Pb^{2+} ion (as well as around Tl^+ ion in KCl-Tl). Vale [240] also came to the same conclusion. Study of KBr and KI crystals doped with thallium and indium (the concentration of 5 10^{23} m^{-3}) showed that under X-irradiation the decay of anion exciton occurs with the formation of Frenkel pair mainly near the impurity ions.

The decomposition of the solid solution of KBr doped with lithium, sodium, strontium and calcium contributes to the creation $(Br_2)_2$ centers [241]. This is explained by the fact that the bond between the halogen ions is weakened in the areas with high impurity concentrations that promote the formation of atomic halogen.

4.2.3. Effect of Non-Homological Anions

Melik-Gaikazjan *et al.* [242] have shown that the addition of S^{2-} ion in KCl leads to a decrease in the F center accumulation under room temperature irradiation. The authors of this work explain the observed effect following circumstance. Introduction of the divalent anion impurity into AHC lattice leads to the appearance of the same number of anionic vacancies. According to the law of mass action, increase in the concentration of anion vacancies causes a decrease in the concentration of cation vacancies. The lowering in the number of cation vacancies leads to a decrease in the number of H centers localized in the cation vacancies. At the same time, it should be noted that the effect of S^{2-} ions observed in the first stage of accumulation, when the F centers are formed in before-irradiation anionic vacancies which should be greater in doped crystal than in the pure crystal. A similar effect was observed at low temperatures by the authors of [243].

On the contrary, oxygen or hydride ion lowers the radiation resistance of AHC thereby accelerating significantly the coagulation of F centers [244]. X-irradiation of NaCl-O^{2-} at room temperature leads to a significant increase in the F center formation in the first stage of colouring. The increase of oxygen concentration and quenching of specimens promote accumulation of F centers. Prolonged irradiation causes the intense radiolysis and the eduction of chlorine in NaCl-O^{2-} at temperature above 300 K even by a small dose of ionizing radiation (0.1-1.0 Gy).

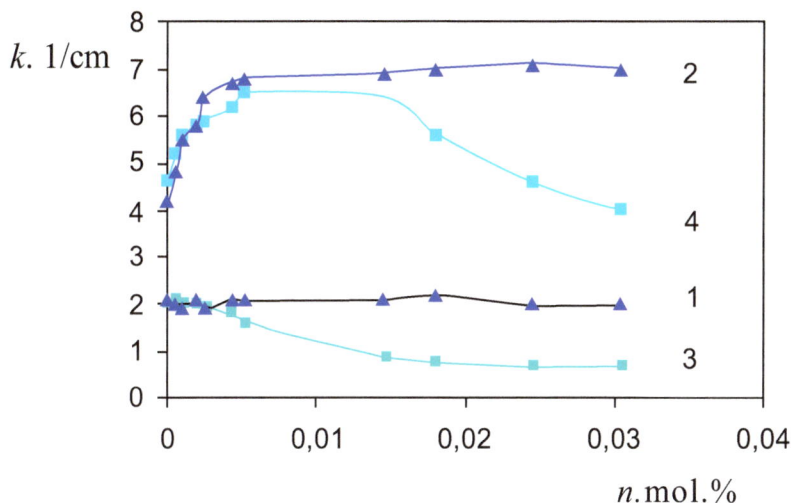

Fig. (4.2). The dependence of the absorption coefficient in the F-band (k) on the lead content (n) in KCl with different histories. 1, 2: aged, 3, 4: hardened (quenched); 1,3: 360 Gy, 2,4: 2 10^4 Gy [234].

It should be noted that doping of the small amount of oxygen (less than 5 10^{-3}) inhibits the radiation colouring. The latter is connected with the formation of chemical compounds of oxygen with uncontrolled impurities. These chemical reactions lead to the purification of crystals and reduce the radiation defect formation.

Effect of acid residues on radiation defect formation in AHC was not practically investigated. According to our data, the additive of SO_4^{2-} anion leads to a small increase in the F center formation of in AHC. Fig. (**4.3**) presents the F center accumulation curves for KBr-SO_4^{2-} with different concentrations of impurity.

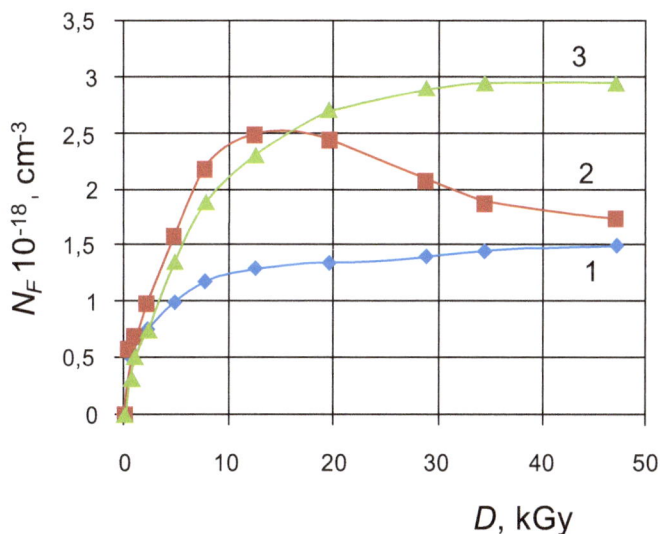

Fig. (4.3). F center accumulation curves for KBr and KBr-SO_4^{2-}. 1: KBr; 2: KBr-SO_4^{2-}, concentration of SO_4^{2-} is 6 10^{-4} mol %; 3: KBr-SO_4^{2-}, concentration of SO_4^{2-} is 7.3 10^{-3} mol %.

In case of smaller concentration of an impurity, the accumulation curve passes through a maximum (12 kGy). A rather low effect of the sulfate-ion addition is apparently connected with a small solubility of this impurity. Quenching of the samples almost does not affect the efficiency of F center accumulation, but leads to a shift of maximum of dose depending for the less doped sample in the direction of lower doses (4 kGy).

CHAPTER 5

Diffusion Under Irradiation

Abstract: Some experimental data on the radiation-stimulated diffusion in metals and alkali halides are presented. The diffusion mechanisms in the crystal solids under irradiation and the effect of temperature irradiation on the diffusion are discussed. It is noted that the super-equilibrium point defects play a major role in the radiation-stimulated diffusion.

Keywords: Radiation, metal, alkali halide, diffusion, diffusion mechanisms, point defects, radiation jolting, implantation.

The change of structure in the solid is not possible without the movement of atoms. Therefore, the radiation-induced diffusion, along with radiation defect formation, is a fundamental phenomenon in irradiated solid solutions.

In the literature, there are a large amount of data to measure the diffusion coefficient in metals and alloys under irradiation with various types of radiation. The observed enhancement of diffusion has a very wide range from 1 to 10^{15} or more [245]. The highest values of radiation effects were obtained at low temperatures (close to room temperature and below) after irradiation by proton, α-particles or ions. The activation energy of diffusion in some cases greatly decreases during irradiation, and sometimes reaches zero.

Obtaining reliable data on radiation-stimulated diffusion requires certain experimental conditions. The absorbed radiation distribution should be uniform, the radiation intensity must be sufficient to generate a steady concentration of defects exceeding thermal equilibrium concentration at irradiation temperature. On the other hand, the radiation power and the absorbed dose should not lead to a noticeable change in the structure of the samples (*e.g.*, the formation of a significant additional number of sinks for radiation defects). At low radiation effects and for large concentration gradients of diffusant, it is necessary to have very precise methods for analyzing the distribution of impurities in the sample.

5.1. EXPERIMENTAL DATA FOR METALS

Consider a few examples of experimental studies of diffusion under irradiation.

Smirnov *et al.* [246] have executed direct measurements of radiation-induced self-diffusion in α-iron. Measurements were carried out in fast neutron flux (ϕ = 1.3-1.4 10^{16} m^{-2} s^{-1}, Φ = 3.5-5.7 10^{21} m^{-2}) in the temperature range of 150-450 °C. The results showed a significant increase in the radiation diffusion compared with thermal diffusion. At the same time, the activation energy of the radiation-induced self-diffusion (0.77 eV) was equal to the activation energy of vacancy migration in iron without irradiation. This indicates that the radiative increase of the diffusion coefficient is related with the generation of super-equilibrium vacancy concentration.

In some cases, the radiation-accelerated diffusion depends on the alloy composition. For example, the diffusion coefficient of nickel in pure nickel is three times higher than in the alloy Fe-20Cr-20Ni and ten times more than in the Fe-20Cr-60Ni under irradiation with 300 keV nickel ions up to ϕ = 1.2 10^{-4} dpa s^{-1} [247]. This shows that in the alloy enriched with nickel, the radiative diffusion coefficient of nickel is several times lower than in the alloy with lower nickel content.

Bystrov *et al.* [248] studied the radiation-enhanced diffusion in the Ag-8.75at.%Zn alloy irradiated with 2.3 MeV electrons (ϕ = 2 10^{17} m^{-2} s^{-1}) in the temperature range from -20 to +190 °C. Based on these results, they drew the following conclusions: (i) In the case of annihilation of point defects at dislocations (unsaturated sinks) the diffusion coefficient does not depend on temperature and is proportional to the flux of electrons. (ii) If the radiation defects disappear by mutual annihilation, the activation energy is $E_{mi}/2$ for diffusion-controlled migration of interstitial atoms, or $E_{mv} - E_{mi}/2$, if a vacancy diffusion mechanism is determining. In this case, D is proportional $\sim \phi^{1/2}$. (iii) If the radiation defects disappear both on dislocations and by mutual annihilation, the dependence of the diffusion coefficient on temperature and on the flux of electrons has a complicated form. Experimental data show that D^{irr} is exponentially dependent on temperature and is proportional to the square root of the flux of electrons. This proves that the mutual annihilation of radiation defects dominates in the experimental conditions applied. The experimental value of diffusion activation energy during irradiation is equal to 0.41 eV that differs from the value for both the vacancy (0.53 eV) and the interstitial (0.1 eV) diffusion mechanism.

Acker *et al.* [249] found that under irradiation of extremely high neutron flux intensity ($\phi = 1.5 \ 10^8$ dpa s^{-1}) at 35 ° C, the diffusion coefficient of copper and gold in aluminum increases by 10^6 times. But at 220° C, the radiation effect is absent. It is assumed that the interstitial contribute is more than the vacancies in radiative enhancement of diffusion. To explain the observed effect Acker *et al.* attracts also the diffusion mechanism involving di-vacancies.

From the results of investigation of radiation-enhanced diffusion in metals follows that in the process of mass transfer during irradiation have participate the recoil atoms, vacancies, crowdions, mono and double interstitial atoms have participated. According to the literature, several mechanisms of radiation-stimulated diffusion are discussed.

5.2. BALLISTIC DIFFUSION MECHANISM

Ballistic mechanism (otherwise recoil mechanism, displacement mixing) is to move the diffusant only due to the energy of incident particles. Movement by this mechanism weakly depends on temperature, but strongly depends on the stopping power. Fig. (**5.1**) shows experimental dependence of the self-diffusion on temperature in nickel under nickel ion irradiation, according to Naundorf [250]. Independence of the diffusion coefficient at low temperatures arises mainly due to ballistic transport. The temperature dependences of the diffusion coefficient for a nickel based alloy during irradiation (displacement rate of 10^{-6} dpa s^{-1}) are shown in Fig. (**5.2**) for various sink densities. This data was calculated from rate theory [251].

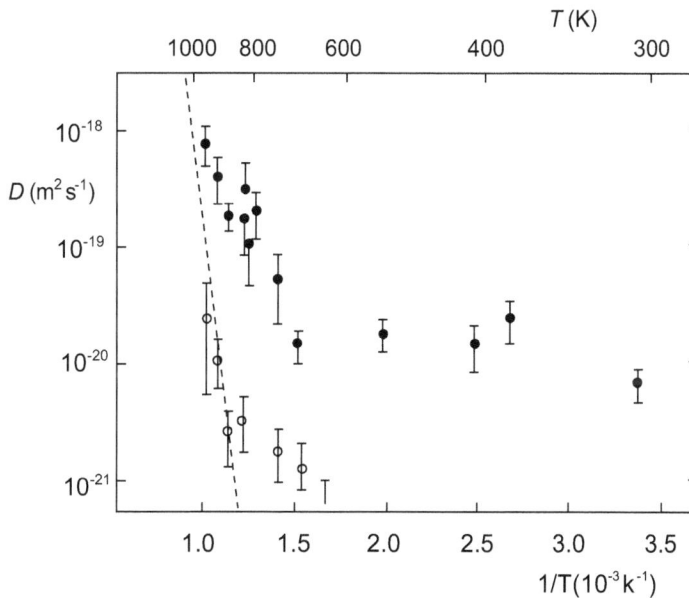

Fig. (5.1): Arrhenius plot of the self-diffusion coefficient in nickel under self-ion irradiation with displacement rates of 1.2 10^{-2} dpa s^{-1} (●) and 1.2 10^{-4} dpa s^{-1} (○). Thermal self-diffusion in nickel is indicated by the dashed line [250].

According to Wiedersich [251], the diffusion coefficient resulting from displacement mixing can be written as

$$D_{mix} = C\frac{\lambda^2}{6}K \cdot \tag{5.1}$$

Here C is the constant, λ^2 is the mean square displacement distance, K is the displacement rate.

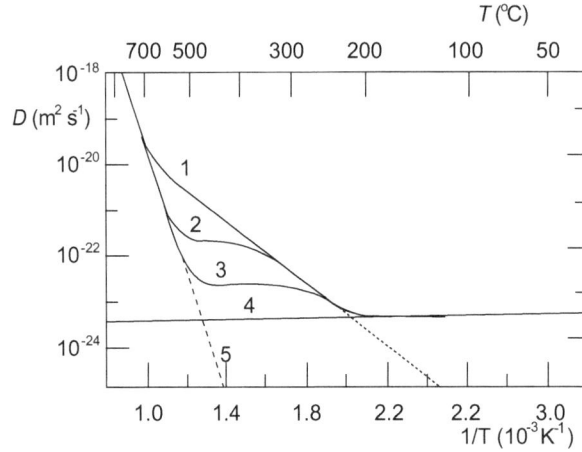

Fig. (5.2): Self-diffusion coefficient in Ni as a function of temperature for different sink densities (ρ_s) under irradiation with $\phi = 10^{-6}$ dpa s^{-1} [251]. 1: no sinks, 2: $\rho_s = 10^{-4}$, 3: $\rho_s = 10^{-3}$, 4: ballistic mechanism, 5: thermal diffusion coefficient.

In the study of radiation stability of solid solutions with phase precipitates under the high-energy irradiation, it is important to take account of the efficiency of ballistic mechanism during the atom transport across the interface. The corresponding measurements have been performed for the α-β interface in Cu-Pd alloy and for the γ-γ' interface in Ni-Mo PE16 alloy [252]. The result is 1-2 nm^2 dpa^{-1}.

5.3. POINT DEFECT DIFFUSION MECHANISM

Mass transfer in crystalline solids is carried out mainly by vacancy or interstitial mechanisms. The first one occurs through the exchange of places between atom and vacancy. As it is known, the diffusion coefficient for any mechanism can be written as [253]:

$$D = D_0 \exp\left(-\frac{E_m}{kT}\right),$$ (5.2)

where $D_0 = \nu \varpi$, E_m is the migration energy, ν is the oscillation frequency of the diffusant, ϖ is the probability of diffusant jump (the probability of free position finding in the nearest environment). From Eq. (5.2) it is apparent that the radiation can affect the diffusion coefficient through three factors: ν, ϖ and E_m. In the case of vacancy diffusion mechanism ϖ is proportional to the vacancy concentration. Therefore, if the steady-state concentration of radiation vacancies exceeds the concentration of thermal vacancies, irradiation should increase the efficiency of the transport. If the transfer occurs with the participation of interstitials, the diffusion coefficient is proportional to the interstitial concentration. In general, the transfer can be carried out by both vacancies and interstitials.

Sizmann [254], Lam *et al.* [255] and others under modeling of radiation-enhanced diffusion took into account of the transfer by both types of diffusion mechanisms. Under irradiation the diffusion coefficient is the sum of two terms: interstitial diffusion coefficient (D_i) and vacancy diffusion coefficient (D_v):

$$D^r = D_i n_i + D_v n_v.$$ (5.3)

Here D_i and D_v are interstitial and vacancy diffusion coefficients, respectively, n_i and n_v are the concentrations of interstitial atoms and vacancies, respectively. The stationary concentration of point defects is determined by their mutual recombination and trapping at sinks. This situation is described by the following rate equations:

$$\frac{\partial n_v}{\partial t} = K - k_{iv} n_i n_v - k_{vs} n_v n_s + D_v \Delta n_v,$$ (5.4)

$$\frac{\partial n_i}{\partial t} = K - k_{iv} n_i n_v - k_{is} n_i n_s + D_i \Delta n_i,$$ (5.5)

where K is the defect formation rate, n_s is the sink concentration, Δn_v and Δn_i are the concentration gradients of point defects. Sometimes, the last terms in the right side of Eqs. (5.4 and 5.5) are not counted. The rate constant for the mutual recombination is:

$$k_{iv} = \frac{4\pi r_{iv}}{\Omega}(D_i + D_v), \tag{5.6}$$

Rate constants for annihilation of interstitials or vacancies, respectively, at the fixed sinks are:

$$k_{si} = \frac{4\pi r_{si}}{\Omega}D_i, \tag{5.7}$$

$$k_{sv} = \frac{4\pi r_{sv}}{\Omega}D_v. \tag{5.8}$$

Here r_{iv}, r_{sv}, r_{si} are the capture radius of the vacancy for an interstitial and of a sink for a vacancy or interstitial, respectively, Ω is the atomic volume.

Depending on the irradiation conditions (temperature, effective volume of the defect capture and sink concentration) two fundamentally different cases are possible. The situation in which most of the interstitials and vacancies disappear due to mutual recombination, and the steady-state concentration is determined precisely by this process called "mutual or pair recombination". In this case the diffusion coefficient (D_r^r) is proportional to the square root of the defect production rate.

$$D_r^r \cong a_0^2 f\left(\frac{K v_i}{\alpha}\right)^{1/2}. \tag{5.9}$$

If the bulk of the interstitial atoms disappears as a result of entering on sinks and the steady-state concentration of vacancies is much higher than the concentrations of interstitial atoms (due to lower mobility of vacancies), then D^r should vary linearly with the defect production rate. This case known as "fixed annealing on sinks" or "fixed vacancies", and is described by expression:

$$D_s^r \cong a_0^2 f\frac{K}{\beta n_s + \alpha n_v}. \tag{5.10}$$

Here a_0 is the lattice parameter, f is the correlation factor (for interstitial diffusion $f = 1$),

$$v_i = v_{i0} \exp\left(-\frac{E_{mi}}{kT}\right), \tag{5.11}$$

E_{mi} is the migration energy of self-interstitials, α is the recombination volume, β is the sink capture volume. The results of the study of radiation-enhanced self-diffusion in iron which are represented by Smirnov *et al.* [246] can be an example of the case of fixed vacancies.

Schüle and Kornmann [256] by means of the considered model have analyzed the experimental data on the nickel diffusion in nickel under 2 MeV electron irradiation in the temperature range between 550-800 K, and have come to a conclusion that radiation-enhanced diffusion coefficient of nickel are described by Eq. (5.9). The best agreement of calculation results with experimental data occurs at the following parameters: $K = 2 \cdot 10^{-8}$ s^{-1}, $n_s = 10^{-11}$, $v_{i0} = 6 \cdot 10^{13}$ s^{-1}, $E_{mi} = 1.0$ eV. Herewith the experimental value of diffusion activation energy equals to 0.82 eV that has appeared to be below the accepted value of 1.0 eV. Here lowering the activation energy is explained by the attraction of interstitial atoms to vacancies. Calculated from the experimental data the recombination volume was found to be 10^5.

From the analysis of experimental data on radiation-induced diffusion obtained by other authors, Schüle and Kornmann, came to the conclusion that the diffusion of gold in gold and gold and copper in aluminum is better described by Eq. (5.9) and silver in the silver by Eq. (5.10). Magnitude of activation energy of radiation-enhanced diffusion is closer to the values of the diffusion by the interstitial mechanism than by a vacancy (see Table **5.1**). This is evidence that diffusion in given systems and the accepted irradiation condition occurs by "One or Two Interstitial-Model".

Table 5.1. Various Parameters and Experimental Data for Radiation-Enhanced Diffusion in Some Metals [256]

Materials	Irradiation	Flux, $m^{-2} s^{-1}$	K, dpa s^{-1}	E_{mi}, eV	E^r, eV
Ni in Ni	electrons	$1.25\ 10^{19}$	$2\ 10^{-8}$	1.0	0.82
Au in Au	neutrons	$6\ 10^{16}$	$2\ 10^{-8}$	0.80	0.58
Cu in Al	neutrons	$6\ 10^{16}$	$5\ 10^{-8}$	0.58	0.40
Au in Al	neutrons	$6\ 10^{16}$	$5\ 10^{-8}$	0.58	0.40
Ag in Ag	protons	$1.25\ 10^{17}$	$1.6\ 10^{-6}$	0.67	0.50

5.4. DI-VACANCY DIFFUSION MECHANISM

It is obvious that the diffusion mechanism implemented by a jump in the double vacancy is easier than by a jump in single vacancy. In usual conditions but not at very high temperatures, the concentration of di-vacancies is very small ($n_{2v} \sim n_v^2$) and the transport mechanism of di-vacancy can be neglected. Under irradiation the vacancy concentration can strongly increase. In connection with this Kiv *et al.* [257] supposed that in some cases di-vacancies play the main role in the radiation-induced diffusion. For Al-Si alloy the process of hopping of the silicon atom into mono and di-vacancy has been thoroughly reviewed. A silicon atom was placed in the center of the calculation cell, which consisted of 102 atoms. Calculations have shown that for a jump of Si atom into a di-vacancy it is necessary to overcome a potential barrier, whose value is an order of magnitude smaller than for mono-vacancy ($E_{mSi}^{2v} = 0.052$ eV and $E_{mSi}^{v} = 0.57$ eV, accordingly). A similar calculation for the self-diffusion in aluminum gave the following values: $E_{mAl}^{2v} = 0.08$ eV and $E_{mAl}^{v} = 0.32$ eV. The concentration of mono and di-vacancies was assumed to be equal to the sum of heat and radiation-induced vacancies:

$$n_{v(2v)} = n_{v(2v)}^{th} + n_{v(2v)}^{irr}.$$ (5.12)

In steady state the radiation-induced concentration of di-vacancies was determined by the expression:

$$n_{2v}^{irr} = \phi\sigma N \exp\left(\frac{E_{mAl}^{2v}}{kT}\right).$$ (5.13)

Here ϕ is the neutron flux, σ is the effective cross-section for mono-vacancy creation, N is number of Al atoms per unit volume. Since the vacancy concentration depends on the temperature, the diffusion coefficient under irradiation has been calculated for three temperature ranges:

(i) High-temperature mono-vacancy diffusion (200-600 °C)

$$D = D_{0h} \exp\left(\frac{-E_{mSi}^{1v} - E_f^{1v} + E_{mAl}^{1v} + E_b^{1v}}{kT}\right).$$ (5.14)

(ii) Intermediate-temperature mono-vacancy diffusion (50-200 °C)

$$D = D_{0i} \exp\left(\frac{-E_{mSi}^{1v} + E_{mAl}^{1v} + E_b^{1v}}{kT}\right).$$ (5.15)

(iii) Low-temperature di-vacancy diffusion (20-50 °C)

$$D = D_{0l} \exp\left(\frac{-E_{mSi}^{2v} + E_{mAl}^{2v} + E_b^{2v}}{kT}\right).$$

(5.16)

Here D_{0h}, D_{0i} and D_{0l} are the pre-exponential factors. D_{0i} is proportional to the concentration of radiation mono-vacancies, D_{0l} is proportional to the concentration of radiation di-vacancies. E_f^{lv} is the formation energy of mono-vacancy, E_b^{lv} and E_b^{2v} are the binding energies of silicon with a mono-vacancy and di-vacancy, respectively. Effective activation energy of silicon diffusion in different temperature regions are as follows: $E_h = +0.81$ eV, $E_i = +0.05$ eV and $E_l = -0.053$ eV. Dependence of radiation-stimulated diffusion on temperature is shown in Fig. (5.3). Comparison of calculation results with experimental data shows that di-vacancy diffusion model of silicon in aluminum is a good description of the diffusion-controlled processes in the alloy Al-Si at low temperatures.

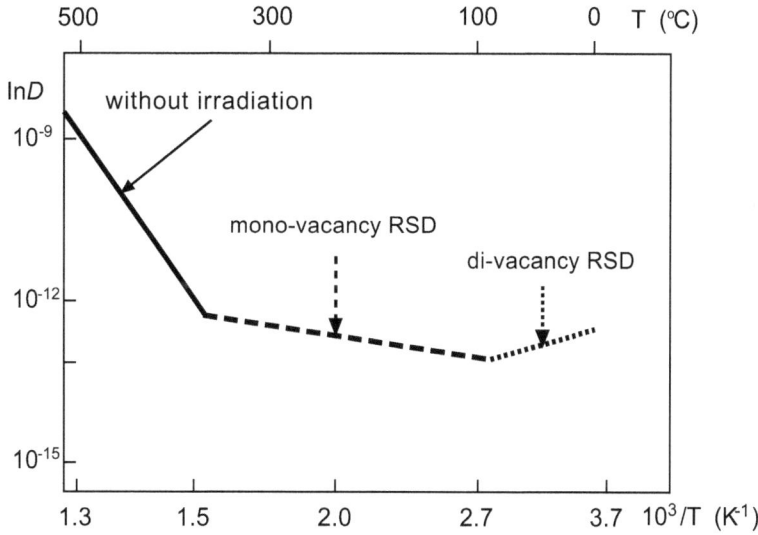

Fig. (5.3). Temperature dependence of Si radiation-stimulated diffusion (RSD) under irradiation in Al-Si alloy [257].

5.5. DIFFUSION UNDER IMPLANTATION

Data on radiation-accelerated diffusion obtained for the implanted samples has presented in number of researches. Piller and Marwick [47] have simulated the erosion profile of the implanted chromium in nickel irradiated with 100 keV He^+ ions at $\phi = 4\ 10^{17}$ m^{-2} s^{-1}. The flux of impurity atoms is given by:

$$J_a = \varphi D^r \nabla n_a + \sigma n_a J_d,$$

(5.17)

Where $\varphi = D_a^r / D^r$, D^r and D_a^r are the radiation diffusivities of solvent and solute respectively, n_a is the solute concentration, σ is the constant, J_d is the defect flux. Concentration of point defects was calculated by using the equations of chemical kinetics. The coefficient of radiation-enhanced diffusion is defined as

$$\Omega D^{irr} = f_v D_v n_v + f_{2v} D_{2v} n_{2v} + f_i D_i n_i.$$

(5.18)

Here f_v, f_{2v} and f_i are the correlation factors for vacancies, di-vacancies and interstitials, respectively. The calculations well describe a distribution profile of the chrome implanted at a room temperature after an irradiation with helium ions at 500 °C.

A similar experiment was conducted by Piller and Marwick on Ni-Cr alloy doped silicon. It has appeared that the presence of silicon significantly reduces the mobility of chromium. This effect is not due to the interaction of Cr with Si but is associated with a decrease in the concentration of point defects because of their recombination near the silicon. To explore the radiation-enhanced diffusion in the presence of point-defect traps, a model has been used which is expressed in usual chemical kinetics equations. This adopted model takes into account the recombination of point defects near the impurity, loss of defects to dislocations and thermal production of vacancies. The initial values of parameters were chosen so as to correspond to the conditions of present experiments. The calculations showed that the degree of radiation effect on the transport of chromium in Ni-0.33at.%Si strongly depends on the binding energy of the impurity-radiation defect (E_b). The value of $D_v n_v$ is given as

$$\frac{1}{6}\left[\frac{2\lambda^2 K}{n_a} + \lambda^2 v_v \exp\left(-\frac{E_b}{kT}\right)\right].$$

(5.19)

Here λ is the unit cell edge length (0.3 nm), K is the defect formation rate, v_v is the frequency with which the vacancy jumps between neighboring sites. The expression (5.19) shows that radiation-enhanced diffusion is temperature independent if the second term is much smaller than the first, *i.e.* at lower temperature and bigger value of E_b. Results of calculations are presented in Fig. (**5.4**).

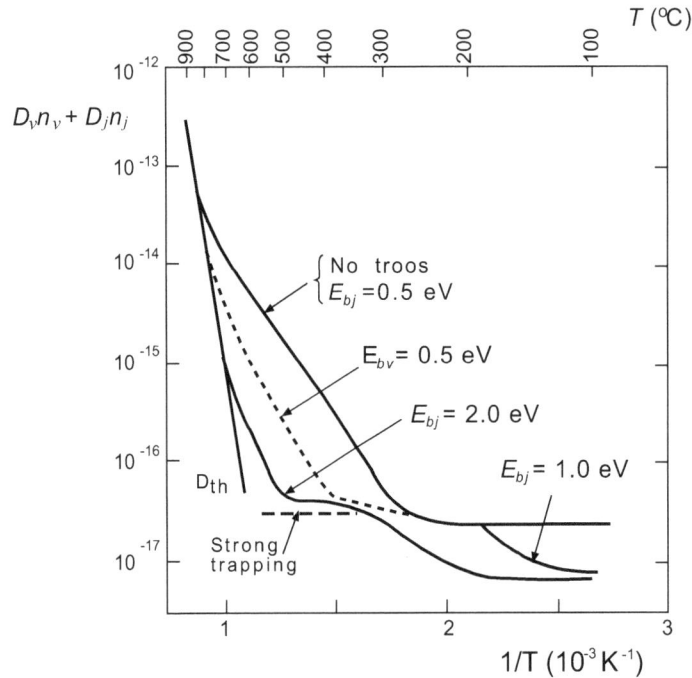

Fig. (5.4). Results of model calculations of solvent diffusion coefficient under irradiation in the presence of point-defect traps. E_{bi} and E_{bv} are binding energies of solute-interstitial and solute-vacancy, respectively [47].

It should be noted that in [47] data the profiles of implanted samples after annealing at high temperatures without irradiation are not presented.

5.6. INTERSTITIAL DIFFUSION MECHANISM

In the literature there is extremely little data on the effect of radiation on interstitial diffusion mechanism. On the basis of experimental data on research of the radiation-accelerated ordering of a silver-zinc alloy Bystrov *et al.* [258] offered the following expression for the radiation-enhanced diffusion coefficient of interstitial:

$$D_i^r = \frac{K}{\alpha\sqrt{\dfrac{8\pi r_v K}{\lambda^2 \alpha}t + 1}}.$$

(5.20)

Here K is the defect formation rate, α is the dislocation density, r_v is the capture radius of interstitial by vacancy, λ is the jump distance, t is the irradiation time.

5.7. THE THERMODYNAMIC APPROACH TO DESCRIBE THE DIFFUSION DURING IRRADIATION

Kirihara [259] used a thermodynamic approach for the analysis of experimental data on radiation-induced self-diffusion in gold under neutron irradiation. Generation of point defects causes a change in free energy ΔG, which is associated with the change of entropy (ΔS^{irr}) and enthalpy (ΔH^{irr}) of system. Here ΔS^{irr} is the sum of changes in configuration (ΔS_c^{irr}) and vibrational (ΔS_v^{irr}) entropy, and ΔH^{irr} is the sum of changes in the enthalpy of formation (ΔH_f^{irr}) and migration (ΔH_m^{irr}) vacancies. Since the lattice constant increases under an irradiation, ΔS_v and ΔH_m should decrease. According to the laws of thermodynamics we obtain the following relation:

$$\Delta G = \Delta H^{irr} - T\Delta S^{irr}. \tag{5.21}$$

It is known that

$$\ln D = \ln(\nu\lambda^2) + \frac{S}{k} - \frac{H}{kT}, \tag{5.22}$$

where ν is the jump frequency, λ is the jump distance, S and H are entropy and enthalpy without irradiation. Because

$$\ln D^r = \ln(\nu a^2) + \frac{S - \Delta S^{irr}}{k} - \frac{H - \Delta H^{irr}}{kT}, \tag{5.23}$$

it is possible to write down

$$\ln D^{irr} - \ln D = \frac{\Delta G}{kT} \tag{5.24}$$

and

$$\frac{\Delta G}{kT} = -\frac{\Delta S^{irr}}{k} + \frac{\Delta H^{irr}}{kT}. \tag{5.25}$$

The values of $\Delta G/kT$, $\Delta S^{irr}/k$ and $\Delta H^{irr}/kT$ can be obtained from the experimental data. Fig. (**5.5**) shows the schematic temperature dependence of radiation-enhanced diffusion for different ratios between ΔH^{irr} and H (**a**), and experimental dependences for self-diffusion in gold under and without neutron irradiated (**b**). The experimental data are also used to construct the dependence $\Delta G/kT$ vs. $1/T$ which is shown in Fig. (**5.6**). Calculation by Eq. (5.25) gave the following results: $\Delta S^{irr}/k = 18.8$, $\Delta H^{irr} = 1.24$ eV, $T_c = 766$ K. $T_c = \Delta H^{irr}/\Delta S^{irr}$ that corresponds to a condition when the work done by the system is zero.

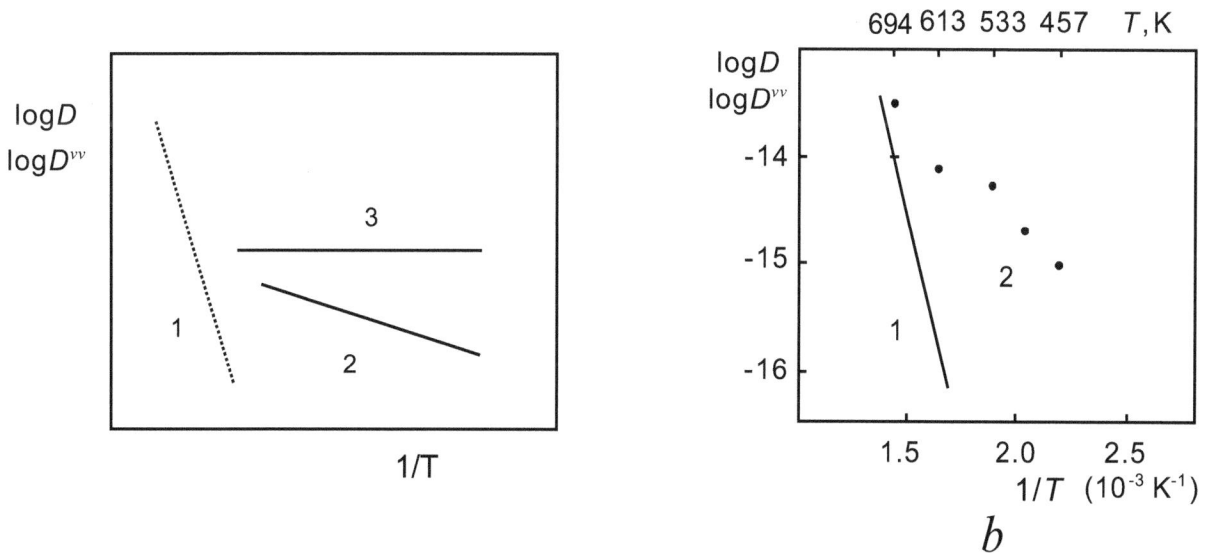

Fig. (5.5). *a* – Schematic presentation of the enhanced diffusion coefficient D^{irr} vs. $1/T$, 1: for thermal diffusion, 2: $\Delta H^{irr} < H$, 3: $\Delta H^{irr} = H$; *b* – self-diffusion of Au (experimental data), 1: without irradiation, 2: under irradiation. [259].

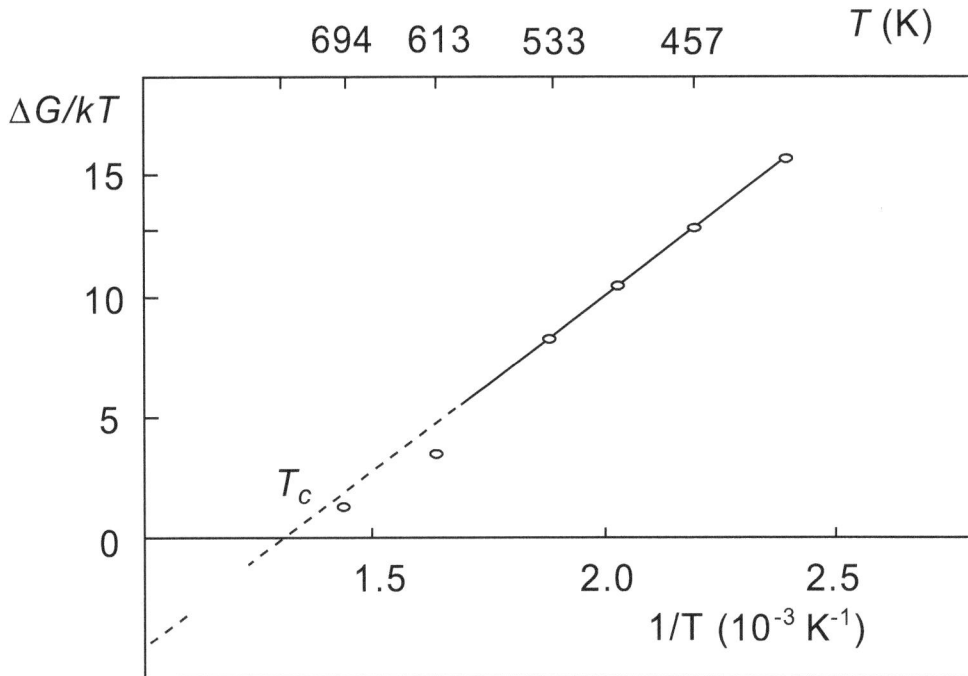

Fig. (5.6). $\Delta G/kT$ vs. $1/T$ calculated from self-diffusion coefficient of Au under neutron irradiation.

5.8. RESULTS FOR DIFFUSION IN IRRADIATED AHCs

Data on radiation-accelerated diffusion for AHCs in the literature is presented considerably less than for metals. Most of the findings to increase the transfer rate in AHCs under irradiation are made on the basis of indirect data on the decomposition of solid solution. For example, Lajzan *et al.* [260] found that gamma-irradiation ($Pd \approx 3$ Gy s^{-1}) accelerates several times the association of single manganese ions in sodium chloride at room temperature. Similar results were obtained in [261-263].

Direct measurements of the radiation diffusion coefficient of ^{22}Na in KCl were performed by Gotlib *et al.* [264]. They found that electron irradiation at 333 K does not lead to faster Na diffusion even at doses of 10^8 Gy. At the same time, the diffusion coefficient of Cl under electron irradiation to dose of 10^7 Gy at room temperature has the same value as without irradiation at a temperature of 600 °C. The increase in the radiation migration of chlorine ions is caused by the formation of interstitial atoms and ions of the halogen (H and I centers), which have a very high mobility.

Annenkov and Galanov [265] found an increase in the diffusion of ^{22}Na in NaCl, NaBr and NaCl$_{0.9}$Br$_{0.1}$ under electron irradiation. The activation energy ^{22}Na was equal to 0.07-0.09 eV in the temperature range of 300-673 K.

Zakhryapin *et al.* [266] studied the diffusion of ^{204}Tl isotopes in KCl, KBr and CsI in the field of γ-radiation ($P_d = 1.5$ Gy s^{-1}, $D \approx 10^5$ Gy). It appeared that in the temperature range of 350-550 °C, the irradiation does not affect the diffusion of thallium in investigated alkali halides. Absence of radiation effect is connected, apparently, with a relatively low dose rate at which the steady-state concentration of radiation defects was lower than the thermal concentration.

Surzhikov *et al.* [267] studied the diffusion of Mg^{2+} and Al^{3+} cations in KBr under 1.4 MeV electron irradiation. Dose rate was 1.5-2.7 kGy s^{-1}. The diffusant depth distribution was measured by using a mass spectrometric method after annealing the irradiated and unirradiated samples in the temperature range of 673-923 K. The values of diffusion coefficients of aluminum and magnesium without irradiation and with irradiation, respectively, were as follows:

$$D_{Al} = 1.1 \cdot 10^{-13} \exp\left(-\frac{0.61}{kT}\right), \ D_{Al}^{r} = 1.7 \cdot 10^{-9} \exp\left(-\frac{1.23}{kT}\right);$$

$$D_{Mg} = 3 \cdot 10^{-15} \exp\left(-\frac{0.39}{kT}\right), \ D_{Mg}^{r} = 1.3 \cdot 10^{-12} \exp\left(-\frac{0.71}{kT}\right).$$

From the data presented it is visible that the irradiation leads to an increase in the activation energy of jump and in the D_0. The authors explain the increase of activation energy as a result of increase in the size diffusant because of capture of one and more electrons by Al and Mg ions during irradiation. Raising the D_0 is due to a significant increase in the concentration of cation vacancies generated by a powerful electron beam.

5.9. RADIATION JOLTING

The interesting mechanism of radiation effects on defects in crystalline solids, called as "radiation jolting", was proposed by Indenbom [268]. According to the proposed model, the unstable Frenkel pairs recombine and thereby cause a slow decaying with the distance of longitudinal waves, which put pressure on any object. Under the influence of such pressure, the object can athermal move. The magnitude of pressure can be estimated by the equation:

$$p = \frac{1}{6\pi}\frac{\chi}{Rv^2}\Omega_F''(t),\tag{5.26}$$

where χ is the modulus of dilatation, R is the distance from annihilated Frenkel pairs, v is the velocity of longitudinal waves, Ω_F is the surplus volume of Frenkel pair.

To confirm the mechanism of radiation jolting, Aluker *et al.* [269] investigated the formation of F centers and a luminescence of self-trapped excitons in pure KCl and KBr, and in same crystals doped with Na, Li and I. The samples were irradiated with pulses of electrons (400 keV, 1 kA, 3 ns). The authors consider this result as a confirmation of the "radiation jolting".

CHAPTER 6

Decomposition of the Solid Solutions Under Irradiation

Abstract: The experimental data of the effect of irradiation on the decomposition of alloys and doped alkali halide crystals are presented. Depending on the composition of the solid solution and other conditions, the irradiation can both accelerate and induce the decomposition. The mechanisms and some features of the decomposition of solid solution under irradiation are discussed.

Keywords: Radiation, metal, alloys, alkali halide, solid solution, decomposition of solid solution, precipitation, precipitates.

Irradiation of any substance leads to excitation of the electronic subsystem and the formation of structural defects. This impact increases the internal energy of the system and makes it unstable. In accordance with the laws of thermodynamics, the processes decreasing the internal energy and leading to the equilibrium condition must occur in such system. However, since the considering system is an open system, therefore, the equilibrium condition cannot occur in it. In cases when the processes of irradiation energy dissipation stabilize, a stationary condition of the system under irradiation can be considered. The stability of solid solution in usual conditions is determined by the interrelation of real concentration of the impurity (n) and the solubility limit (n^{th}). The unsaturated solution ($n < n^{th}$) is stable and the supersaturated one ($n > n^{th}$) is unstable and must decompose. The decomposition of solid solution is realized through accumulation of impurity in the form of homogenous and/or heterogeneous precipitates. The decomposition rate depends on the mobility of impurity atoms. At low temperatures, when the diffusion of impurity is "frozen", the supersaturated solution may stay in the unstable condition as long as desired. Under irradiation, the radiation damages occurring in the solid solution will impact on the factors determining the stability of solid solution, namely on the solubility limit and mobility of impurity atoms. Therefore, one of the important questions originated while researching the radiation-stimulated decomposition of solid solution is the question whether the impurity precipitation is radiation induced or radiation accelerated. In the first case, exposure to radiation creates a new or an additional driving force for solution decomposition. In the second case, exposure to radiation accelerates the spontaneous process of impurity precipitation. Fulfillment of the first and the second cases is supported by observed data.

6.1. RADIATION-ENHANCED DECOMPOSITION

Radiative acceleration of decomposition of the solid solution is possible only in supersaturated solutions, i.e., when the concentration of single impurity atoms exceeds the solubility limit.

The example of radiation-enhanced decomposition of the supersaturated solid solution is an origin of copper precipitates in the iron-copper alloy during its exposure to radiation by fast neutrons at the temperature of 290 °C [270]. To decrease the concentration of single atoms of copper two-fold in the alloy with the initial concentration of copper $n_{0Cu} = 0.14$ at. % the absorbed dose of 0.5 dpa is required, and for $n_{0Cu} = 0.08$ at. % 0.18 dpa is required (solubility limit is 0.026 at. %). The comparison of calculation results with the observation data provides a possibility to state that the main role in the nucleation of copper clusters is played by cascade collisions.

150 keV proton irradiation of Cu-12.4at.%Be alloys at the room temperature results in the formation of Guinier-Preston zones [271], that is the decomposition of solid solution through spinodal mechanism. A complete decomposition is observed for irradiation doses at $3 \cdot 10^{-3}$ dpa. Same effect without an irradiation is obtained after a thermal treatment at the temperature of 200 °C within the period of 65 hours. The observed radiation enhancement can be explained by the vacancy mechanism for the Be transport.

Radiative acceleration of decomposition can be observed in a very wide temperature range. Shriver and Richardson [272] in the course of researches of radiation impact on the micro-hardness of nickel-carbon alloy, determined that exposure to radiation by fast neutrons ($\Phi = 10^{22}$ m^{-2}) intensifies the decomposition of supersaturated solid solution Ni-C in the range of temperatures between 20-1200 °C. The detected effect is connected with interaction of carbon atoms with vacancies, and it may contribute to the formation of carbon or carbide (Ni_3C) precipitates. Irradiation at the temperatures above 500 °C results in the change of form and stability of precipitates or other traps for the complexes of carbon-vacancy.

Enhancement of decomposition occurs not only for an irradiation neutrons and ions, but also for ionizing radiation. For examples, Sagaradze *at al.* [273] determined that electron irradiation (5 MeV, $\Phi = 2 \cdot 10^{22}$ m^{-2}) even at elevated temperatures (500 °C) speeds up the process of Fe-Cr decomposition. Bystrov *et al.* [274] found that irradiation by 2.5 MeV electrons ($\phi = 1.83 \cdot 10^{17}$ m^{-2} s^{-1}) at the room temperature accelerates the decomposition of Cu-2.5%Be alloy significantly.

Poerschke and Wollenberger [275] explored the impact of 3MeV electrons on the Cu-Ni alloys. The alloys with interrelation of components 50:50 demonstrated that their ray treatment by electrons in the whole range of temperatures between 120 and 350 K, stimulates the decomposition of solid solution. At the same time, if the temperature grows, the speed of decomposition increases. The observed decomposition is interpreted by interstitial diffusion. The calculated energy of activation in the radiation-stimulated decomposition was 0.2 eV. The increased value of activation energy in relation to the activation energy of interstitial atoms in the pure metals (0.1 eV) is explained by contribution of formation energy of the interstitial (0.4 eV) to the energy of movement activation. If the temperatures are higher, the radiation processes will be determined by the vacancy diffusion.

Researching the decomposition of supersaturated alloy Ni-14.3at.%Be in the course of its 2.5 MeV electron irradiation at the temperature of 180 °C [254], it was determined that the maximum of the distribution curve of impurity precipitate volume moves to the side of smaller concentrations of beryllium with the fluence growth. It is necessary to note that if the temperature is higher than 140 °C the speed of solute precipitation does not depend on the temperature and it is proportional to the defect-formation rate, and if the temperature is lower than 140 °C then it becomes slower and it is proportional to $K^{\approx 0.7}$.

The work [276] demonstrates that during the thermal treatment of the Al-Mg-Sc alloy at the temperature of 350 °C, electron irradiation results in the deepening of solid solution decomposition. In the alloys which were annealed at the temperature of 150 °C, the irradiation does not have an effect on the decomposition of solution. It may be explained by the fact that the heating at the temperature of 350 °C results in annealing of sinks for radiation defects, and in the samples that were annealed at the temperature of 150 °C, a large number of sinks is reserved. On these sinks the intensive recombination of radiation defects takes place.

Radiation-enhanced decomposition is also observed in high-doped AHCs. Several works [229, 230, 277, 278] demonstrate that irradiation by ionizing radiation at the room temperature significantly increases the speed of lead aggregation in the halides of potassium and sodium. It is determined that even a small dose rate of γ ray (0.17 R s^{-1} in KBr-Pb [229], 0.1 Gy s^{-1} in KBr-Sr [279]) significantly increases the speed of impurity aggregation at the room temperature.

Decomposition of KCl-Eu ($n_{Eu} = 3 \cdot 10^{-4}$) has its own peculiarity. At the beginning of irradiation (10 kGy), the metastable precipitates of europium occur in the crystals KCl-Eu, and if the doses are large (200 kGy), the radiation stimulates the formation of stable impurity precipitates [280]. γ-irradiation ($P_d = 278$ R s^{-1}) of KCl-Eu at the temperature of 100 °C stimulates the decomposition of solid solution accompanied with precipitation of europium in the form of Suzuki phase [281]. The next annealing at the temperature of 240 °C results in precipitation of the $EuCl_2$ phase.

6.2. RADIATION-INDUCED DECOMPOSITION

The demonstrative example of radiation-induced decomposition is presented by Agrault [282] using the Ti-6Al-4V alloy. The sample of such alloy was exposed to radiation by V^+ ions at the temperature of 660 °C within the period of 3 hours. The analysis showed that the new phase precipitates appeared in the sample. Then, the sample was kept in normal conditions at the temperature of 650 °C within the same period of time, as a result, the part of these precipitates dissolved. It is the evidence of the fact that in this case irradiation induces precipitation of the new phase.

The exploration performed by Cauvin and Martin [283] using the transmission electron microscopy accompanied by simultaneous ray treatment in the column of microscope by 1MeV electrons of the solid solution Al-1,9%Zn, demonstrates the formation of Zn precipitates occurs under irradiation at the temperatures which are significantly higher than the temperature of solvus without any radiation. The value of temperature shift depends on the dose rate. This example of the radiation-induced precipitation demonstrates unexpected features: (i) there is no correlation between the place of precipitation and the place of point defect sinks; (ii) precipitation of incoherent β phase occurs with the smaller atomic volume than the atomic volume of matrix, as well as coherent Guinier-Preston zones; (iii) the size of the incoherent β precipitates saturates at large dose. There was suggested a common mechanism of fluctuations in the solute concentration under irradiation, which explains well the coherent Guinier-Preston zones and the creation of solute clusters with more complex structures. The calculations by using the Russell's model for the growth of incoherent precipitates demonstrate that this model may explain qualitatively the observed behavior of the β phase precipitate.

One of the mechanisms of the radiation induction of solid solution decomposition is the ***creation of centers for origination of new phase***. The confirmation of such mechanism is given in the work of Pareige *et al.* [284]. The low supersaturated Fe-Cu (n_{Cu} = 0.088 at. %) alloy was irradiated by 3MeV electrons ($\phi = 10^{17}$ m^{-2} s^{-1}, $\Phi = 10^{23}$ m^{-2}) or 150 keV Fe$^+$ ions ($\phi = 5 \ 10^{14}$ m^{-2} s^{-1}, $\Phi = 0.85 \ 10^{16}$-125 10^{16} m^{-2}) at the temperature of 573 K. After the irradiation by electrons the formation of copper clusters was not observed. Under irradiation by iron ions the clusters of copper appeared only after 100 s of exposition. After an 840 s irradiation the concentration of copper clusters amounted to 10^{24} m^{-3} with the average radius of 1 nm. During the following irradiation the size of copper clusters was growing. The number of copper atoms which were included in the clusters changed from 5 to 70. This dependence of copper cluster formation from the irradiation time testifies that origination of clusters occurs only in the associates of point defects (vacancies or interstitial atoms).

Bakay *et al.* [285] analyze the possibility of radiation formation of nucleation centers. According to their data, irradiation stimulates generation of small precipitates beyond the extended defects, and it is connected with the increase in the speed of nucleation in the displacement cascades, that is a result of significant local changes in the concentration of component solution.

The radiation induction of solid solution decomposition is also found in the AHC. Rubio *at al.* [286] demonstrated that the X-irradiation (50 kV, 30 mA) does not accelerate but namely stimulates the decomposition of solid solution NaCl-Mn. Irradiation of crystals at the room temperature results in the formation of the Suzuki phase, and exposure without any irradiation within the same time period does not result in the change of the NaCl-Mn solution. This proved that radiation induces but do not stimulate precipitation of second phase in the AHCs. The authors of work [286] believe that they are the first who has proved radiation-induced decomposition in AHCs.

More complicated processes take place in the potassium chloride and bromide and sodium chloride and bromide doped by divalent cations (Pb^{2+}, Eu^{2+}) under X-irradiation [287, 288]. It was determined that at the beginning the irradiation results in dissolution of metastable phase, the decrease in concentration of single dipoles M$^{2+}V_c$ and in the growth of number of impurity associates with small sizes, and the growth of impurity phase. Under following irradiation the growth of only impurity phase was observed with a sacrifice in all other impurity defects. The same results were received for the KI-Eu crystal [289]. The radiation acceleration of the dipoles M^{2+}-V_c diffusion in the AHC was also established.

The irradiation of solid solutions in some cases may not induce but ***prevent the decomposition***. Melikhov *et al.* [290] found that irradiation of the quenched Ti-48%Al-2%Nb alloy by the Ti$^+$ ions (60 keV) slows down the processes of phase rearrangement and stabilizes the structure of alloy at the beginning stages of ageing at a temperature of 650 °C.

6.3. EFFECT OF RADIATION ON THE SIZE OF PRECIPITATES

Irradiation of the solid solution, which already contains the solute precipitates, can cause both their growth and dissolution. The direction and magnitude of the effect depend on the prehistory of the sample and the irradiation conditions.

Example of radiation-induced growth of precipitates is ***radiation-enhanced coarsening***. This phenomenon is observed, for example, in the Al-Ge alloy under irradiation by the 200 keV Al^+ ions [291]. After irradiation at the temperature of -100 °C and during warming up to the room temperature, the separate precipitates of Ge are transformed into the group precipitates. It was also determined that during irradiation by ions the dissolution of precipitates by means of cascade mixing takes place. A lowest limit for the recoil dissolution rate has been estimated from the results as $1.2 \cdot 10^{20}$ m^{-3} s^{-1}.

Potter and McCormick [292] presented the results of their study of the redistribution process in the sizes of particles of the γ'-phase (Ni_3Al) in the Ni-12.8at.%Al alloy, under irradiation by the 3.5 MeV Ni^+ ions to the dose of 12 dpa ($\phi = 2.3 \cdot 10^{16}$ m^{-2} s^{-1}). During irradiation at the temperature range of 450-700 °C the growth of particle sizes of Ni_3Al takes place accompanied by the attainment of saturation. If the irradiation temperature grows, the attainment of saturation takes place in case of larger doses. The temperature dependence of the irradiation-enhanced coarsening rate constants showed a plateau at 0.028 nm^3 s^{-1} for irradiation temperatures from 500 to 590 °C. The observed radiation-stimulated phenomenon of coarsening is described well by the theory of coarsening from Lifschitz-Slyozov-Wagner, i.e. by the diffusion-controlled growth, according to Gibbs-Thomson, if we replace the rate constant for thermal coarsening k_{th} by the radiation rate constant [293]:

$$k_{rad} = k_{th}\left(\frac{n_v}{n_{vth}} + \frac{n_i D_i}{D_a} + \frac{n_c D_c}{D_a}\right), \tag{6.1}$$

where n_v, n_i, and n_c are concentrations of vacancies, interstitials and complexes of point defects, respectively, n_{vth} is a thermally equilibrium concentration of vacancies, D_i, D_a, and D_c are coefficients of diffusion of the interstitial, solute and complex, respectively .

The magnitude of radiation impact strongly depends on the composition of the solid solution and the type of radiation. When the Ni-12.8at.%Al and Ni-12.7at.%Si alloys are irradiated by the 3 MeV Ni^+ ions at the temperature of 500 °C, we may observe the cubic dependence of precipitate sizes from the dose [293]. At the same time, the irradiation of the austenitic stainless steel by 0.4 MeV Al^+ ions ($\phi = 10^3$ dpa s^{-1}) up to the dose of 100 dpa does not result in the change of size and morphology of precipitates at the temperature of 600 °C [294].

In other investigations a decrease has been observed in the precipitate size under the influence of radiation. Wollenberger [295] found that irradiation of the Cu-1.4at.%Be dilute alloy by Cu^+ ions results not only in the segregation at the dislocation loops in accordance with the interstitial mechanism, but also in the formation of γ-phase precipitates in the areas of alloy enriched with beryllium. The size of precipitates depends on the defect-formation rate (see Fig. (**6.1**)). If defect-formation rate grows, the precipitate size decreases. It shows that the size of precipitates is defined by the competition of processes connected with the diffusion growth and the dissolution which takes place in accordance with the ballistic mechanism.

Wanderka *et al.* [296] explored the effect of irradiation by ions 300 keV Cu^+ on the stability of chrome precipitates in the Cu-1.15at.%Cr-0.03at%Zn-0.14at.%Si alloy. The precipitates of chrome were decreasing in size in all irradiation conditions (80-883 K, $\phi = 1.16 \cdot 10^{14}$ and $1.16 \cdot 10^{17}$ m^{-2} s^{-1} = $2.3 \cdot 10^{-5}$ and $2.3 \cdot 10^{-2}$ dpa s^{-1}, $\Phi = 0.01$ and 20 dpa). After irradiation at the room temperature with the dose of 0.1 dpa ($\phi = 2.3 \cdot 10^{-5}$ dpa s^{-1}) the precipitates became invisible. At the temperature of 803 K and the dose of 20 dpa the precipitates decreased only on 50 % in sizes.

The irradiation of the Ti-6Al-4V alloy by the 9MeV electrons with the dose of 2 dpa stimulates the precipitation of solute β-phase. At the same time, when the irradiation temperature is decreased the density of precipitates grows up, and their size decreases [297].

The interesting data about the influence of irradiation on the precipitates of zinc in the alloy Al-11.8Zn are presented in the work [298]. The irradiation by the 2 MeV electrons ($\phi = 10^{18}$ m^{-2} s^{-1}) at the temperature of 320 °C at the beginning causes the decrease in number of zinc atoms in the precipitate, then the size of precipitates grows up. As we can see from Fig. (**6.2**) the minimal size of precipitate is observed with the dose of $\Phi \approx 1.2 \cdot 10^{21}$ m^{-2}. The growth of precipitates in case of big doses is explained by the authors of the work [298] as compared with the growth of large precipitates by means of dissolution of small ones. The possible reasons of precipitate dissolution in cases of small doses are not discussed.

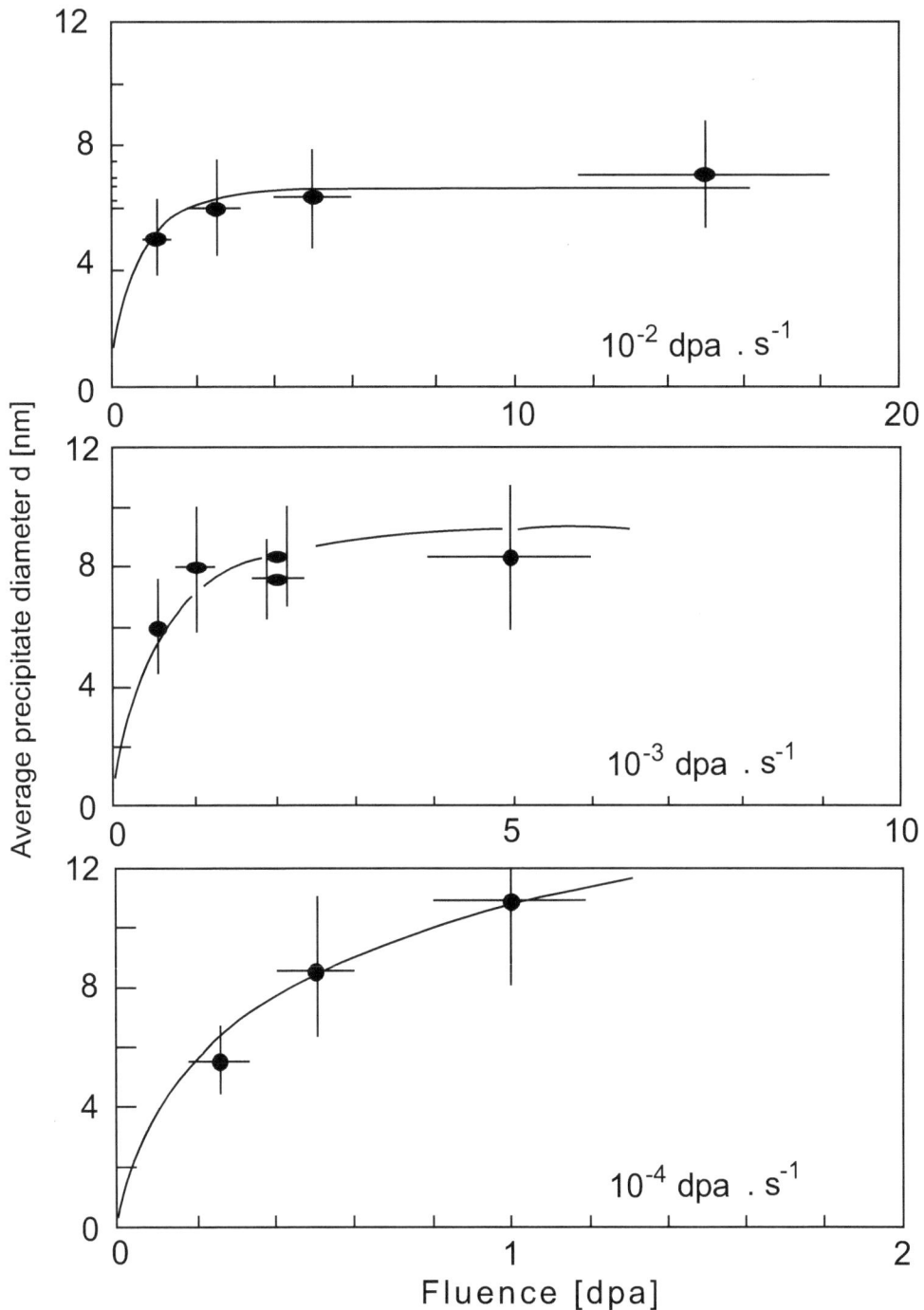

Fig. (6.1). Fluence dependence of the average γ precipitate diameter ($\varnothing$) for Cu-1.4at.%Be irradiated by 300 keV Cu$^+$ at 600 K [295].

Potter and Hoff [299] researched the impacts of irradiation by the 3.5 MeV Ni$^+$ ions on the solute precipitates in the Ni-6.35wt.%Al alloy. The electron microscopy measurements demonstrated that during irradiation at the temperature of 550 °C the medium size of γ'-phase (Ni$_3$Al) grows from 4 to 12 nm at the beginning, and then decreases, and in cases of maximum doses ($\varPhi \approx 35$ dpa) it reaches the value of 8.5 nm. The initial growth of precipitate size is conditioned by the radiation-stimulated coarsening, and the following decrease is explained by the interaction of precipitates with dislocations and dislocation loops generated by the radiation.

6.4. HOMOGENEOUS DECOMPOSITION

Radiation-stimulated homogeneous decomposition is typical for solid solutions with high solubility, or for dilute solutions.

When the Ni-41.4at.%Cu alloy is irradiated by the 3MeV electrons ($\phi = 7 \ 10^{13}$ cm^{-2} s^{-1}, $\Phi = 0.66$-$5.8 \ 10^{19}$ cm^{-2}) in the temperature range of 373-510 K, the randomly distributed small areas enriched by one or other component are formed [300]. The observed coherent precipitates are specific for the spinodal decomposition of solid solution. The peak intensity of concentration fluctuations increases with increasing irradiation dose. The average cluster-cluster distance is of the order of about 2.5 nm.

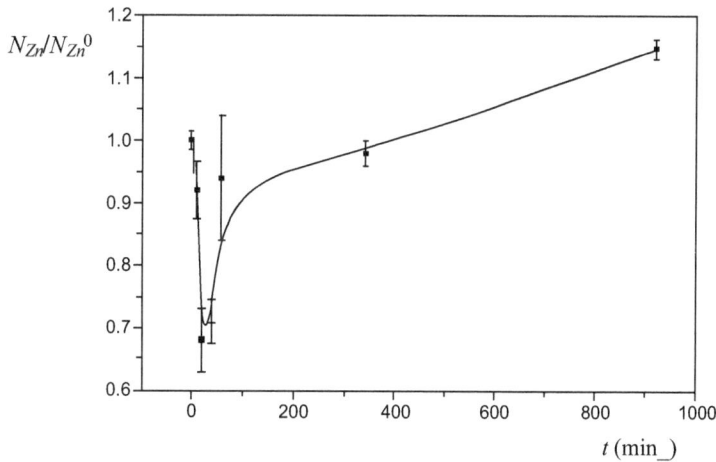

Fig. (6.2). Number of Zn atoms in precipitate (normalized to the initial value) as a function of 2 MeV electron irradiation time [299].

Mukai and Mitchell [301] explored the radiation-stimulated decomposition of the unsaturated solid solution (Ni-1at.%Be). At the beginning of irradiation by the 650 keV electrons ($\phi = 5 \ 10^{22}$ m^{-2} s^{-1}), the decomposition passed in accordance with the homogeneous mechanism. In this case irradiation causes the formation of Guinier-Preston zones and precipitates of the γ''-phase. During the following irradiation the precipitation took place on the dislocation loops of interstitial type. The longer irradiation produces the tangles of dislocations containing precipitates. The received result may be explained by the movement of solute-interstitial complexes to pre-existing solute clusters as well as specimen surfaces.

6.5. OTHER FEATURES OF RADIATION-STIMULATED DECOMPOSITION OF ALLOYS

The interesting phenomenon was observed by Cawthorne and Brown [302] under irradiation of the M316 stainless steel by fast neutrons. Transmission electron microscopy of several samples of M316 type stainless steel after irradiation in the Fast Reactor has revealed a high concentration of small precipitates giving a sublattice type diffraction pattern with the same unit cell and approximately the same lattice parameter as the austenite matrix. The precipitates have observed in the specimen having the lowest irradiation temperature (270 °C) and also in the specimen with the lowest dose (3.2 dpa). The size of precipitates was in the range of 8-23 nm, and the concentration was in the range from $8 \ 10^{20}$ to $4.2 \ 10^{21}$ m^{-3}. The chemical composition was not detected but it was similar to Cu_3Au.

This kind of radiation has a significant impact on the efficiency of nucleation of impurity precipitates. This was detected by Ishino *et al.* [303] exploring the impact of irradiation by the 2 MeV electrons and 100 MeV ions on the electrical resistivity (ρ) of the Fe-0.02wt.%Cu and Fe-0.6wt.%Cu alloys at the temperature of 300 K. Even small values of the dose rate and the absorbed dose ($\Phi \leq 10^{-4}$ dpa) cause the significant change of ρ. In the first moment of irradiation the quick growth of ρ takes place, than it decreases monotonously (see Fig. (**6.3**)). The initial growth of ρ

is connected with origination of copper clusters. Apparently, small embryos increase the real resistivity due to the strain field around them. The decrease of ρ is specified by the depletion of single copper atoms as a result of their aggregation. At the same time, as we can see from the figure, the irradiation by electrons causes a higher speed of ρ change than heavy ion irradiation. The calculations demonstrate [303], that from the 10 Frenkel pairs which were formed by electron irradiation, 16 atoms of copper emerged from the solid solution, and by carbon ion irradiation the number of atoms was several times smaller.

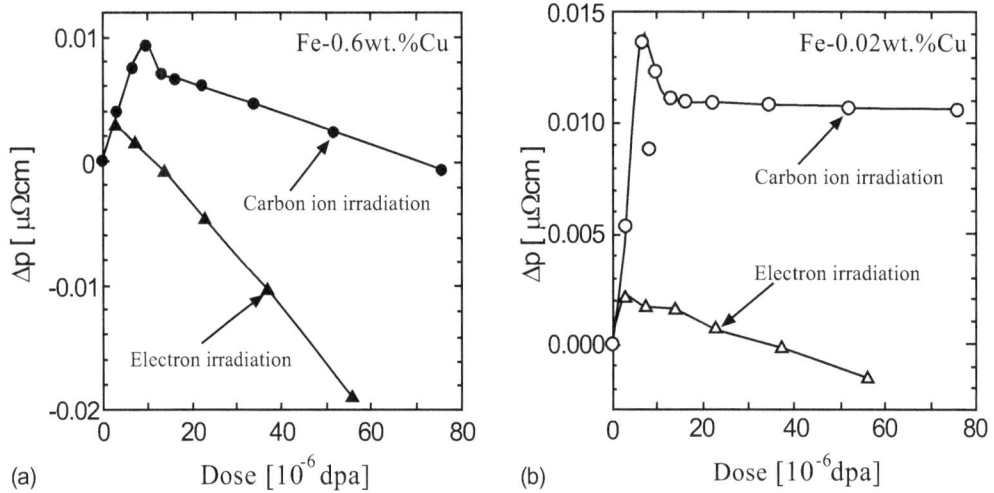

Fig. (6.3). Dpa dependence of electrical resistivity change by irradiations at 300 K with 2 MeV electrons and 100 MeV carbon ions; (**a**) Fe–0.6wt.%Cu alloy, (**b**) Fe–0.02wt.%Cu alloy [303].

The irradiation of steels including the carbon by the reactor neutrons practically always results in the formation of carbides of solute and solvent. The carbides are the most thermally stable from all radiation-induced phase precipitates. Therefore, the irradiation of the austenitic stainless steel at the temperature of about 400 °C results in the formation of niobium carbide and the areas enriched with silicon [304]. The irradiation at the temperature of 600 °C results in formation of phases enriched with silicon and nickel. When the temperatures are above 600 °C the phases enriched by nickel are absent.

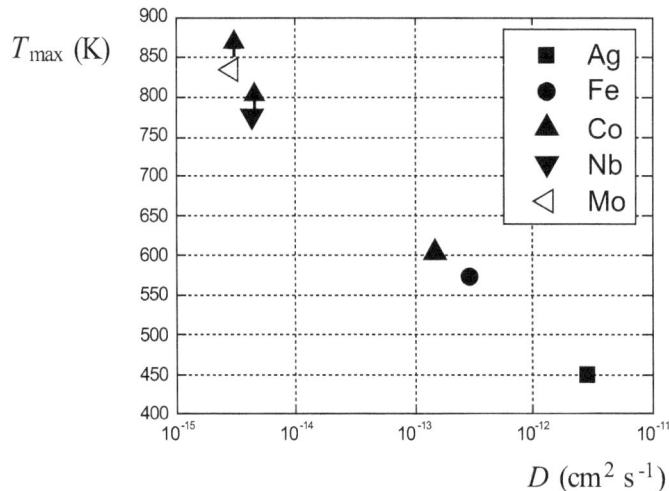

Fig. (6.4). Dependent of the maximum temperature for self-organization during irradiation (T_{max}) with 1.8 MeV Kr vs. the solute diffusion coefficient in Cu at 900 K [306]. For Nb and Mo, macroscopic coarsening was not observed up to the temperatures indicated.

According to the opinion of Koptelov and Subbotin [305], irradiation of supersaturated solid solution by the cascade-forming radiation may result in the specific mechanism of solute redistribution. Due to the significant increase of local temperature during the process of thermalization ($\approx 10^{-11}$ s) the thermo-diffusional movement of solute atoms from the surrounding media into the volume of cascade is possible. At least, the enrichment of cascade volumes is possible for the interstitial atoms.

Chee *et al.* [306] studied effect of 1.8 MeV Kr$^+$ ions bombardment on alloys of copper with the limited solubility in a wide range of temperatures. At lower temperatures most of the alloys formed nearly homogeneous solid solutions. At intermediate temperatures these same alloys underwent macroscopic phase separation. At higher temperatures (above some critical temperature T_{max}) occurs a macroscopic coarsening. It is established that value of T_{max} depends on diffusion coefficient of the solute (see Fig. (**6.4**)). The lower diffusion coefficient, the higher value of T_{max}. Macroscopic coarsening can be suppressed if the ratio of diffusion by thermally activated processes to that by ballistic events (ion beam mixing) kept small. For Nb and Mo, macroscopic coarsening was not observed up to the temperatures indicated.

CHAPTER 7

Theoretical Estimates and Models of the Radiation Stability of Crystal Solid Solutions

Abstract: The factors affecting the stability of solid solutions under irradiation are considered. One of these factors is the radiation-induced change of solubility. The models describing the radiation-induced decomposition of the solid solution for the different concentrations of impurities and irradiation conditions are discussed.

Keywords: Radiation, metal, alloys, alkali halide, solid solution, radiation stability, solubility, precipitation, precipitates, decomposition of solid solution.

7.1. THEORETICAL ESTIMATES OF THE RADIATION STABILITY OF SOLID SOLUTIONS

The simplest and most natural approach to assess the effect of irradiation on phase stability of solid solution is a thermodynamic estimate of the system during irradiation. The thermodynamic approach is to identify changes in Gibbs free energy due to the formation of radiation point defects. The calculations performed by Kovin and Martin for the Al-Zn alloy at the temperature of 235 °C [307] demonstrate that in case of medium dose rate of $\phi = 2 \ 10^{-2}$ dpa s^{-1}, 10^{-5} vacancies and 10^{-8} interstitial atoms are formed. When considering the energy of vacancy formation (0.66 eV) and interstitial atom formation (3.2 eV) we receive $\Delta G = 6.63 \ 10^{-6}$ eV per atom. This change of free energy is not enough for realization of phase transition for which it is necessary to have $\Delta G > 1.1 \ 10^{-3}$ eV per atom.

For the first time the effect of radiation on solid solutions by using the thermodynamic approach was analyzed presumably by Wilkes *et al.* [308]. According to their estimation, in case of ultra equilibrium concentration of vacancies in 10^{-4} and the typical enthalpy of formation 1 eV, the internal energy should be increased upto 10^{-4} eV per atom. This change of internal system energy is close to the energy which is necessary for phase transitions in the metals. They also came to the decision that the radiation-defect formation in the solid solution results in the shift of maximum change of free energy to the side of component which has the largest binding energy. At the same time, the authors of this work [308] highlight that this theory badly conforms to the experiment, and the calculated diagrams of phase equilibriums are the unreliable indicators for the quantitative estimation of a system state under irradiation.

Yamauchi *et al.* [309] evaluated the effect of vacancy excess on the system which is subject to phase transition when the temperature is changed. Their calculations demonstrate that irradiation must decrease the temperature of phase transition.

Bocquet and Martin [310], using the thermodynamic approach estimated the impact of radiation-defect formation on the stability of binary solid solutions (**A**-**B**). They analyzed the ternary solution where the vacancies (*V*) were used as the third component. Due to relatively small concentration the interstitials were not taken into account. In accordance with the chosen model the free energy of solid solution (*G*) is defined by the energy of pair interaction of components of the solution (E_{ii}, E_{jj} and E_{ij}) and by the change of system entropy. E_{ij} is the nearest neighbor pair energy between species *i* and *j* in the ternary solution. Here the indexes *i* and *j* denote two types of atoms in the solution **A**, **B** or *V*. The change of free energy of solution is described by the following expression:

$$\Delta G = z \sum_{i,j} W_{ij} n_i n_j + kT \sum_i n_i \ln n_i, \tag{7.1}$$

$$W_{ij} = E_{ij} - \frac{1}{2}(E_{ii} + E_{jj}), \tag{7.2}$$

where *z* is the coordination number, n_i and n_j are the concentrations of the ternary solution components including the vacancies. The calculation of free energy was executed by solving the system of non-linear equations with the aid of numerical methods. On the basis of these calculations the phase diagrams were constructed. Analysis of the results is carried out in terms of the W_{ij}. For example, if the values $W_{ij} > 0$, then the phase precipitations appear only on increase of the critical mixing temperature. The separating phase will be disordered, but if $W_{AB} > 0$, $W_{AV} > 0$ and $W_{BV} < 0$, then the separating phase will be ordered and enriched by defects. The phase precipitates cannot appear if $W_{BV} > 0$. Bocquet and Martin analyzed 22 possible variants of ternary phase diagrams. By comparing the results with experimental data for Al-Zn alloy, they concluded that the conditions deduced from the thermodynamic approach that are necessary for irradiation induced precipitation, are not satisfied by the experimental cases.

7.2. RADIATION EFFECT ON THE SOLUBILITY

The driving force of the solid solution decomposition under irradiation and without it is the degree of real concentration excess over the solubility limit. For this reason in the course of radiation-induced decomposition it is necessary to consider that the irradiation decreases the limit of solubility. There are direct evidences of the irradiation impact on the solubility limit in the metal alloys.

Cauvin and Martin [307], exploring the radiation-stimulated decomposition of the Al-Zn alloy determined that under irradiation by the 1MeV electrons at the temperature of 235 °C the solubility of zinc decreases more than 10 times. At the same time as we can see from Fig. (**7.1**), the line of solvus has a sharp dependence on the temperature during irradiation. Increase of dose rate increases the radiation effect.

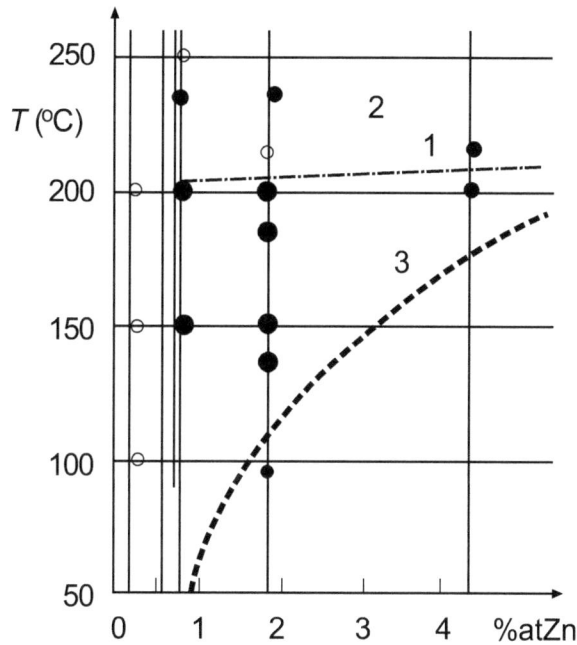

Fig. (7.1). Zn solvus in Al under 1 MeV electron irradiation [307]. Half-circles to the right (left) point to experimental data at high (low) irradiation flux. Full (open) symbols point to the occurrence (absence) of radiation-induced precipitation. 1: 2.5 10^{23} m^{-2} s^{-1}; 2: 2.5 10^{24} m^{-2} s^{-1}; 3: without irradiation.

Using the complex model of radiation stability of solid solutions, Cauvin and Martin [311] received the following expression for the solubility limit under irradiation:

$$n^{irr} = n^{th} B \,. \tag{7.3}$$

Here B is the parameter which depends on the solute diffusion coefficient of interstitial (D_a^i) and vacancy (D_a^v) mechanisms and on the probability of capturing solute by the boundary surface of interstitial atoms (ω_i) and vacancies (ω_v).

$$B = \left[1 + \left(\frac{a_i}{a_v} \left(1 - \frac{1}{S_v} \right) \right)^x \right]^{-1} , \tag{7.4}$$

where $S_v = n_v/n_{vth}$, n_{vth} is a thermally equilibrium concentration of vacancies, $x = -1$, if $\omega_i > \omega_v$, $x = +1$, if $\omega_i < \omega_v$, $D_a^i = a_i D_i n_i$, $D_a^v = a_v D_v n_v$, $\omega_i = n_i \exp(-E_i/kT)$, $\omega_v = n_v \exp(-E_v/kT)$, n_i and n_v are the concentrations of interstitials and vacancies, E_i and E_v are the binding energies of interstitial and vacancy with the precipitate, respectively.

Urban and Martin [312] through the modeling of radiation-induced coarsening, received the following relation for solubility under irradiation:

$$n_a^{irr} = \frac{D_v}{D_v + D_i} n_{ath}.$$ (7.5)

Here D_v and D_i are the diffusion coefficients of vacancies and interstitials, respectively, n_{ath} is the limit solubility without irradiation. If the solid solution already has precipitate with the radius r, then the equilibrium concentration for such particles under irradiation is defined by the expression:

$$n_r^{irr} = n^{irr} \exp\left[\frac{2\sigma\Omega}{kTr} + \ln W(r)\right].$$ (7.6)

Here σ is the interfacial-energy, Ω is the atomic volume, k is the Boltzmann constant, T is the absolute temperature.

$$W(m) = \frac{D_v[\rho_v(m) + \rho_m(m)] + D_i[\rho_i(m) + \rho_m(m)]}{D_v \rho_t(m)},$$ (7.7)

where $\rho_v(m)$, $\rho_i(m)$ and $\rho_m(m)$ are the density numbers of precipitates formed by vacancy, interstitial and those and others mechanism, respectively, m is the number of solute atoms into precipitate, $\rho_t(m) = \rho_v(m) + \rho_i(m) + \rho_m(m)$.

Bakay *et al.* [313] developed the complex model of calculation of the equilibrium concentration of binary alloy **A-B** under irradiation. The model considers the following: the formation of coherent and incoherent precipitate, the radiation increase of the solute diffusion coefficient, and the distinction in the mechanisms of diffusion of the solute atoms which have smaller or bigger size in relation to the atoms of matrix. While solving the system of kinetic equations for the stationary concentration of vacancies the following expression was received:

$$n_v^{irr} - n_v^{th} = \frac{1}{2}\left(n_v^{th} + \frac{k^2}{\alpha_r}\right)\sqrt{1 + \frac{4K}{\alpha_r D_v^{eff}}\left(n_v^{th} + \frac{k^2}{\alpha_r}\right)^{-2}} - 1.$$ (7.8)

Here n_v^{th} is the thermally equilibrium concentration of vacancies, k^2 is the power of sinks for pointed defects, α_r is the rate constant of point defect recombination, K is the defect formation rate.

$$D_v^{eff} = d_{Bv}\frac{1 + (\lambda - 1)n_A^{irr}}{1 + (\omega - 1)n_v^{irr}},$$ (7.9)

$\omega = \lambda/\mu$, $\lambda = \zeta d_{Ai}/d_{Bi}$, $\mu = d_{Av}/d_{Bv}$, $\zeta = \exp(E_{B\to A}^i/kT)$, $E_{B\to A}^i$ is the transition energy of the interstitial atom from the sublattice **B** to the **A**, d_{Av}, d_{Bv}, d_{Ai} and d_{Bi} are partial diffusivities for **A** and **B** components by vacancy and interstitial mechanism, respectively. The expression for the stationary concentration under irradiation looks like this:

$$n_A^{irr} = \frac{n_A^{irr}}{1 - n_A^{irr}} = \frac{n_A^{th}}{1 - n_A^{th}}\left[1 + \left(\frac{n_v}{n_v^{th}} - 1\right)\frac{1 + \omega}{1 + (\omega - 1)n_A^{irr}}\right]^{\frac{1-\omega}{1+\omega}}.$$ (7.10)

Here n_A^{th} is the thermally equilibrium solute concentration, n_v is the average concentration of vacancies in volume.

It is proposed that in alloys with the undersize solute $\omega > 1$, and with the oversize solute $\omega < 1$. The calculation was performed for the model alloys Ni-Si and Ni-Al for the following values of parameters: $K = 10^{-6}$ dpa s^{-1}, $k^2 = 10^{10}$ cm^{-2}, $\alpha_r = 10^{17}$ cm^{-2}, $d_{Bv} = 0.13\exp(-1.18\text{eV}/kT)$ cm^2 s^{-1}, $n_v^{th} = \exp(-1.7\text{eV}/kT)$, the migration energy of self-interstitial is 0.15 eV, $n_A^{th} = 0.283\exp(-0.077\text{eV}/kT)$ and $0.255\exp(-0.062\text{eV}/kT)$ for Si and Al, correspondingly. The results are given in Fig. (7.2), where $E_\mu = E_{Bv}^m - E_{Av}^m$ (eV), $E_\lambda = E_{B\to A}^i + E_{Bi}^m - E_{Ai}^m$ (eV). E_μ is the difference in migration energies using the vacation mechanism for the oversize solute and the atom of matrix; E_λ is the difference in migration energies using the interstitial mechanism for the undersize solute and the atom of matrix, plus the difference in energies for interstitial state of the solute and matrix atoms. From the data presented in Fig. (**7.2**) it is clear that the irradiation decreases the solubility limit of the undersize solute (Si in Ni) which moves according to the interstitial mechanism. For the oversize solute (Al in Ni) the irradiation increases the solubility limit.

We [314] estimated the impact of radiation defect formation on the solubility of impurity in the AHCs. Using the thermodynamic approach we received the following expression for the stationary concentration of impurity in the binary solution under irradiation, accompanied by formation of one type of radiation impurity defects:

$$n^{irr} = n_\infty (1 - z n_d) \exp\left(-\frac{z n_d (E_s + E_b)}{kT}\right). \tag{7.11}$$

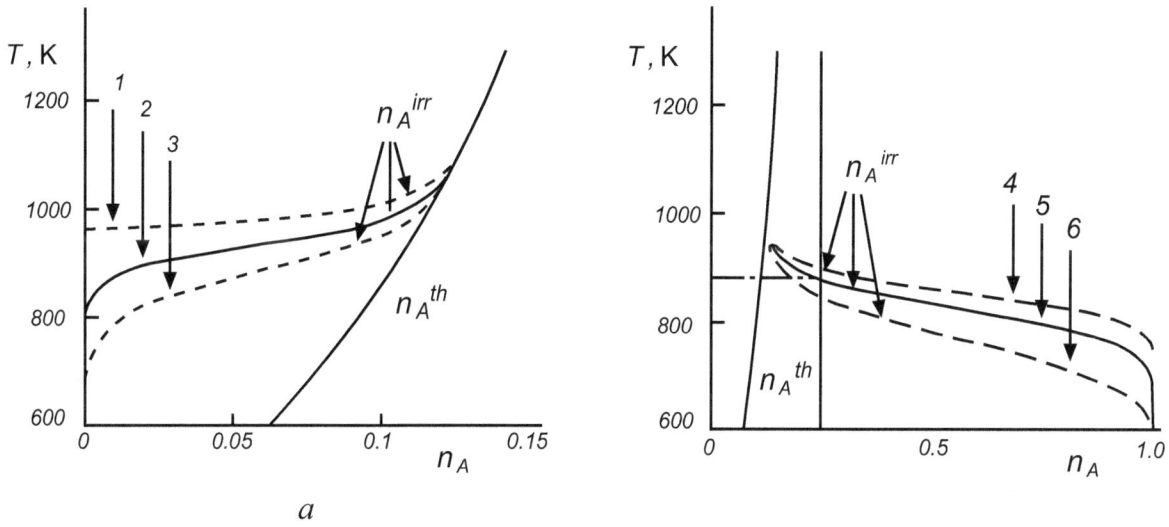

a

Fig. (7.2). Radiation-modified phase diagrams of binary alloys. $K = 10^{-6}$ dpa s^{-1}, **a** – Ni-Si ($\omega > 1$), **b** – Ni-Al ($\omega < 1$). $E_\lambda = 1.0$ eV (1), 0.14 eV (2), 0.05 eV (3), $E_\mu = 1.0$ eV (4), 0.1 eV (5), 0.05 eV (6) [313].

Here z is an effective cross-section of interaction between the radiation defects with the impurity atom, n_d is the concentration of defects, E_s and E_b is mixing enthalpy and the biding energy of defect with the impurity atom, respectively. In the supersaturated solid solutions, the equilibrium concentration of single solute atoms depends on the radius of precipitate (r) and the interface energy (σ) according to the relation:

$$\ln\left(\frac{n_r}{n_\infty}\right) = \frac{2\sigma\Omega}{rkT}, \tag{7.12}$$

where n_r and n_∞ is the equilibrium concentration of impurity for the precipitates of radius r and $r = \infty$. The radiation effect on the equilibrium concentration of impurity may be expressed in terms of radiation temperature changes (ΔT_{irr}) [315]. Then the Eq. (7.12) will be as follows:

$$\ln\frac{n_r^{irr}}{n_\infty^{irr}} = \frac{2\sigma\Omega}{rk(T + \Delta T_{irr})}. \tag{7.13}$$

Taking $n_\infty^{irr} \approx n_\infty$, from the Eq. (7.12) and (7.13) we will receive:

$$\frac{n_r^{irr}}{n_r} = \exp\left(-\frac{2\sigma\Omega\Delta T_{irr}}{rkT(T + \Delta T_{irr})}\right). \tag{7.14}$$

The experimental data (points) and temperature dependences of n_r^{irr}/n_r calculated by Eq. (7.14) for the supersaturated solid solution of KCl-Sn with various values of ΔT_{irr} are presented in Fig. (**7.3**). The $2\sigma\Omega/r$ is actually a binding energy of the impurity atom in the precipitate of radius r. Let this quantity be the sign E_b, than Eq. (7.14) can be rewritten in the following form:

$$\frac{n_r^{irr}}{n_r} = \exp\left(-\frac{E_b}{kT}\frac{\Delta T_{irr}}{T}\right). \tag{7.15}$$

Calculated from Eq. (7.15) value of E_b at $\Delta T_{irr} = 20$ K was found to be 0.11 eV.

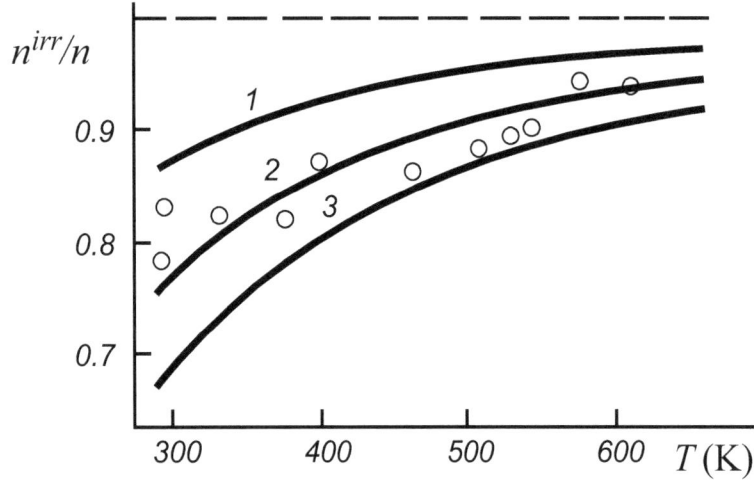

Fig. (7.3). Experimental (points) and calculated by (7.14) temperature dependences of n^{irr}/n for KCl-Sn with different ΔT_{irr}. 1: 10 K, 2: 20 K, 3: 30 K.

More detailed calculation of solubility under irradiation was performed by means of minimization of free energy change ($d\Delta G/dn = 0$) [316]. In the expression for ΔG the binding energy, the energy of thermal decay (E_e) and the concentration (n_c) of the complex impurity-pointed defect, were taken into account.

$$\Delta G = n(n-1)E_s + n_c E_b + T\Delta S, \tag{7.16}$$

where n is the impurity concentration, ΔS is the change of configuration entropy (we neglect the change of vibrational entropy).

$$\Delta S = k[(1-n-n_d+n_c)\ln(1-n-n_d+n_c)+n_c\ln(n_c)+(n_d-n_c)\ln(n_d-n_c)+$$
$$+(n-n_c)\ln(n-n_c), \tag{7.17}$$

$$n_c = \eta n\left[1 - A\exp\left(-\frac{E_e}{kT}\right)\right] \tag{7.18}$$

Here η is the impurity fraction in the complexes, A is a pre-exponential factor. The magnitudes of solubility for KCl-Pb at the temperature of 300 K and $n_d = 10^{-6}$ were received using the numeric computation. The results of calculation (see Fig. (**7.4**)) demonstrated that the radiation effect on solubility initially depends on the value E_b, and on the impurity fraction in the complex. For each system of AHC–impurity there is a critical binding energy of the complex (E_b^*) which weakly depends on the temperature.. The values E_b^* for some doped AHCs are given in the Table **7.1**. If the actual binding energy $E_b < E_b^*$, the radiation defect formation results in the decrease of solubility, if $E_b > E_b^*$, then the solubility increase.

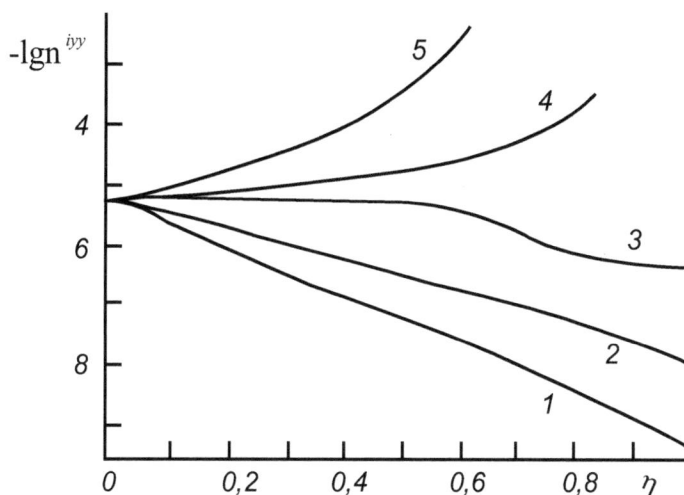

Fig. (7.4). Dependence of lead solubility in KCl on η under irradiation for different binding energies. 1: 0.1 eV; 2: 0.2 eV; 3: 0.3 eV; 4: 0.325 eV; 5: 0.4 eV.

Table 7.1.   Value of E_b^* for $\eta = 0.5$ (eV)

AHC	Eu²⁺	Ba²⁺	Pb²⁺	Na⁺	Cu⁺
KCl	0.34	0.34	0.29	0.22	0.31
KBr	0.34	0.35	0.26	0.20	0.32
KI	0.35	0.28	0.20	0.17	0.35

Fig. (**7.5**) presents a part of phase diagram for KCl-Pb when $\eta = 0.5$ and $n_d = 10^{-6}$. For KCl-Pb $E_b^* = 0.29$ eV. The radiation reduction of solubility is expressed in the line shift of the phase separation to the side of lower concentrations of lead (curve 3).

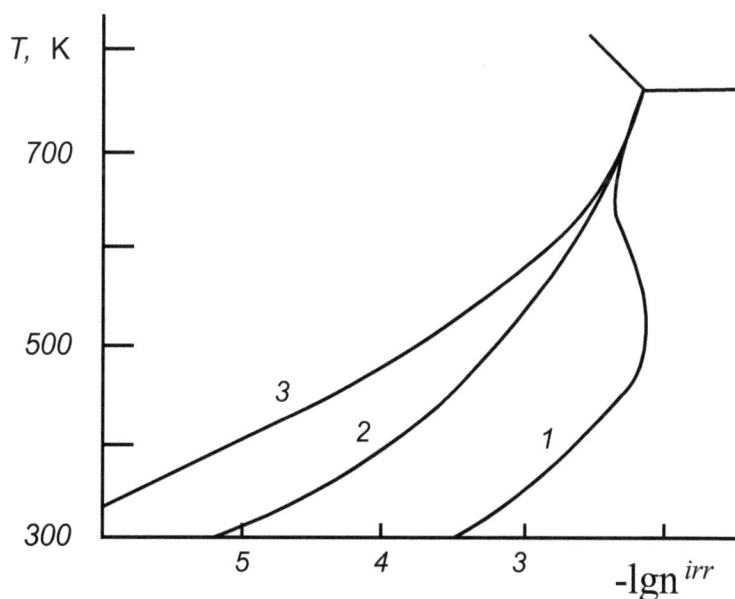

Fig. (7.5) The phase diagram for KCl-Pb for $\eta = 0.5$ and $n_d = 10^{-6}$. 1: $E_b = 0.4$ eV; 2: without irradiation; 3: $E_b = 0.2$ eV.

7.3. KINETIC MODELS OF RADIATION RESISTANCE OF THE SOLID SOLUTION

The kinetic theories of radiation stability of solid solutions turn out to be more fruitful than the thermodynamic. Authors such as Bakay, Martin, Maydet, Nelson, Russell, Urban, Wilkes, and others made great contribution to the theory of radiation resistance of solid solutions. The kinetic analysis of the experiments is based on well known differential equations, using as fundamental parameters for the calculations of the fraction of freely migrating defects, the effective sink concentration for point defect annihilation, the complex formation of intrinsic and extrinsic defects, the motion of solute and solvent atoms by vacancy and interstitial mechanism and *etc*. The results of calculations by the kinetic model both qualitatively and quantitatively describe well the processes of accumulation of radiation defects, the effect of irradiation on the nucleation and growth of coherent and incoherent precipitates, the behavior of impurity precipitates in concentrated and diluted solid solutions under irradiation.

7.3.1. Radiation Growth of Precipitates

Urban and Martin [312] proposed a theory of precipitate coarsening under radiation. The theory is based on the model according to which during irradiation of supersaturated solid solution, the growth of three types of precipitates occurs, and they differ from each other according to the mechanism of solute diffusion (only vacancy, only interstitial and one and another), with which precipitates are formed. The basic equation describing the growth of precipitate under irradiation is based on the well known condition of minimum rate of free energy change. The impact of irradiation on this phenomenon accounted for by introducing a radiation-induced term $B(m)$ (m is the number of solute atoms into precipitate). The theory also considers the change of solubility limit in the solid solution. The factor of radiation impact is presented in the following form:

$$B(m) = \frac{\beta_v(m)[\rho_v(m) + \rho_m(m)] + \beta_i(m)[\rho_i(m) + \rho_m(m)]}{[\beta_v(m) + \beta_i(m)]\rho_i(m)}, \qquad (7.19)$$

where $\beta_v(m)$ and $\beta_i(m)$ are the rates at which solute atoms arrive at a precipitate by the vacancy or the interstitial diffusion mechanism, respectively. Upon performing the necessary transformations the authors of theory received the following equation for solute precipitate growth rate:

$$\frac{dr}{dt} = \frac{\Omega}{r} D_v n_{th} \left[\frac{2\sigma\Omega}{kT}\left(\frac{1}{\bar{r}} - \frac{1}{r}\right) + \frac{<r \ln W(r)>}{\bar{r}} - \ln W(r) \right]. \qquad (7.20)$$

Here $\bar{r}$ is the mean radius of precipitate, $<>$ means average value, the value of $W(m)$ see Eq. (7.7).

The performed numerical calculations with parameters corresponding to the diluted solution Al-Zn providing that $D_i > D_v$, 400 K, $\phi = 0.02$ dpa s^{-1} demonstrated that the irradiation-induced term depends on the radius of precipitate and the interrelation of diffusion coefficients D_i/D_v. The biggest value it has if $r = 0$ (see. Fig. (**7.6**)). With increasing radius of the precipitate, the value of the term decreases and tends to zero. The value of $W(r)$ also depends on the probability of vacancy capture by the precipitate. According to the authors, their theory describes well the rise of coherent precipitates under irradiation.

7.3.2. Unsaturated Solid Solution

Barbu and Martin [317] investigated the precipitation of the Ni$_3$Si phase in the unsaturated solution Ni-4at.%Si. It was found that the irradiation by 2 MeV electrons with the very low dose rate ($5.5 \cdot 10^{-9}$ dpa s^{-1}) in the temperature range of 523-723 K (temperature thresholds) and even with doses of about 10^{-3} dpa, results in the precipitation of solute phase. The received results are described by a simple model, which includes the following suppositions:

(i) The point-defect flux is mainly due to free point-defect migration;

(ii) The solute concentration gradients are sufficiently weak enough to be ignored for the solute concentration dependence of point-defect diffusion coefficient;

(iii) Steady-state conditions prevail.

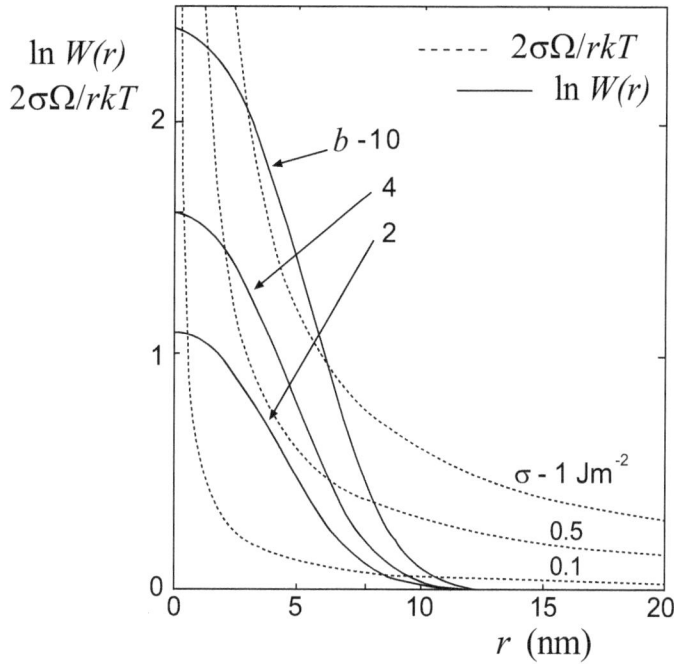

Fig. (7.6). The irradiation-induced term and the interfacial-energy term as the functions of precipitate radius (r) [312]. $b = D_i/D_v$, 1: b =10, 2: b = 4, 3: b = 2, 4: σ = 1 J m^{-2}, 5: = 0.5 J m^{-2}, 6: = 0.1 J m^{-2}.

Solute enrichment factor S_E, i.e., the ratio for the solute concentration at the sink to the average solute concentration in the foil:

$$\frac{1}{S_E} = \left(\frac{a_v}{a_i}\right)^{\alpha} \left\langle \left[\left(1+\frac{a_v}{a_i}\right)S_v - 1\right]^{\alpha}\right\rangle, \tag{7.21}$$

where a_i and a_v are the proportionality factors between the self and impurity-diffusion coefficients by the interstitial and vacancy mechanism, respectively, and

$$\alpha = \frac{b_i + b_v}{a_i + a_v}$$

Here b_i and b_v are the coupling coefficients between the point defect and the solute flux; S_v is the position-dependent vacancy supersaturation between the two sinks and < > stands for the spatial average between two point-defect sinks. The estimation of S_v were obtained by using the standard chemical rate theory for point-defect concentrations under irradiation:

$$\frac{\partial n_v}{\partial t} = K - kn_i n_v - k_v(n_v - n_v^{th}), \tag{7.22}$$

$$\frac{\partial n_i}{\partial t} = K - kn_i n_v - k_i n_i. \tag{7.23}$$

Where n_v^{th} is the vacancy thermal equilibrium concentration, $k_v = D_v\rho_d$ (ρ_d is dislocation density, D_v is the vacancy diffusion coefficient).

7.3.3. Heterogeneous Precipitation

Mruzik and Russell [318] calculated the radiation-stimulated heterogeneous precipitation in the context of Al-Ge and Al-Si alloys. Calculations have shown that supersaturated vacancies produced by irradiation increase the nucleation rate of incoherent precipitates by many orders. The excess of interstitial atoms reduce the nucleation rate, but this effect is smaller than the growing effect resulting from vacancies. The irradiation may also decrease the nucleation rate. But it occurs only in case of very high outputs of irradiation dose rate. The results of calculations correspond well to the observation data for the Al-Ge alloy.

Maydet and Russell [319] proposed a model of spherical incoherent precipitates growth with the relative volume effect $\delta = (\Omega_p - \Omega)/\Omega$ (Ω_p and Ω are atomic volumes of precipitate and matrix, correspondently). The capture of migrant solute atom provides the growth of precipitate, and the capture of vacancies and interstitial atoms compensates the change of volume by decreasing the volume effect for the whole precipitate. The schematic representation of precipitate state in coordinates with the number of solute atoms in the precipitate (m) vs. the concentration of point defects (n_d) is presented in Fig. (7.7). The lines with arrows demonstrate a possible direction of precipitate evolution. The change in the rate of solute number in the precipitate is defined by the equation

$$\frac{\partial m}{\partial t} = \beta_m \left[1 - \exp\left(\frac{1}{kT} \frac{\partial \Delta G}{\partial m} \right) \right],$$ (7.24)

where β_m is the capture rate by precipitate of solute atoms, ΔG is the change of free energy, which depends on the degree of supersaturation ($S_a = n/n_{th}$). There is a critical point which is located at the intersection of nodal lines ($dn_d/dt = 0$ and $dm/dt = 0$). The critical radius corresponds to the critical point

$$r^* = -\frac{2\sigma\Omega}{\Delta\varphi}.$$ (7.25)

Here $\Delta\varphi$ is the radiation-modified potential which depends on the capture rate by precipitate of solute atoms, vacancies and interstitials, on the supersaturation by solute atoms and defects, on the value of Young's modulus. This model is consistent with a number of experimental data, but is not suitable for describing the incoherent precipitates with a negative volume effect.

7.3.4. Effect of Irradiation on the Stability of Precipitate

Nelson *et al.* [320] analyzed two mechanisms of radiation-stimulated dissolution of precipitates. One of them is the recoil dissolution and the other one is the disordering dissolution. In the first case, the rate of dissolution is as follows:

$$\frac{dV}{dt} = -\frac{4\pi r^2 \varphi}{N},$$ (7.26)

where r is the radius of spherical precipitate, φ is the number of displacements per unit time per unit area, N is the number of atoms per unit volume. In the second case, the dissolution rate may be determined by the following equation:

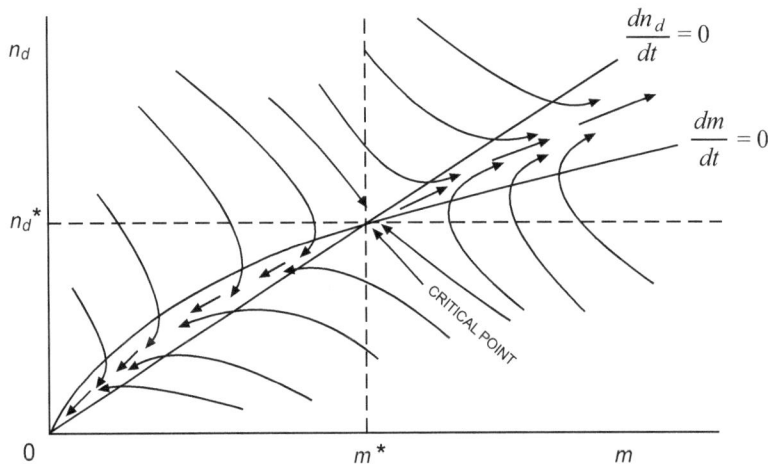

Fig. (7.7). Schematic illustration of nodal lines, critical point, and particle trajectories [319].

$$\frac{dV}{dt} = -4\pi r^2 lfK,$$ (7.27)

where l is a shell of thickness at the precipitate surface which results in the loss of solute atoms by diffusion to the matrix, f is a fraction of solute atoms which actually became dissolved, K is the number of displacements per unit time, per one atom. The rate of precipitate radius changes while dissolving through ballistic mechanisms is given as:

$$\frac{dr}{dt} = -\frac{\phi}{N} + \frac{3(D + D^r)n_0}{4\pi pr} - (D + D^{irr})r^2 m.$$ (7.28)

The rate of the disordering dissolution is given as:

$$\frac{dr}{dt} = -lfK + \frac{3(D + D^r)n_0}{4\pi pr} - (D + D^{irr})r^2 m.$$ (7.29)

Here D and D^{irr} are thermal and radiation-enhanced diffusivities of solute, respectively, n_0 is the total concentration of solute atoms, p is the fraction of solute atoms constituting the precipitate phase, m is the number of precipitates per unit volume.

The stability of precipitates under irradiation depends on the balance between radiation dissolution of precipitates and their growth by enhanced through irradiation and thermal diffusion. There exists a critical temperature below which the dissolution rate exceeds the growth rate (see Fig. (**7.8**)). Since this critical temperature has a place in the regime, when the radiation-enhanced diffusion coefficient is proportional to dose rate, the value of critical temperature will be dose-rate dependent, for example: $T_c \approx 300$ °C for $K = 10^{-2}$ s^{-1}, $T_c \approx 150$ °C for $K = 10^{-6}$ s^{-1}. The dependencies of precipitate radius on the dose for the two dose rates are given in Fig. (**7.9**).

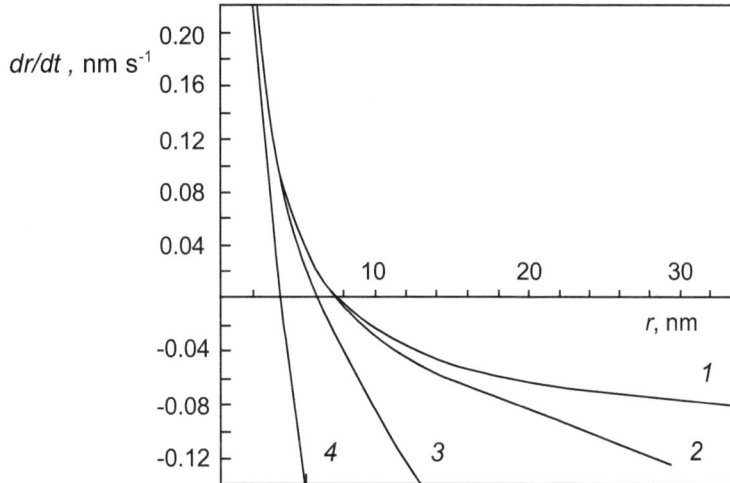

Fig. (7.8). Rate of change of precipitate radius (dr/dt) as a function of radius for different values of m (1: 10^{15}, 2: 10^{16}, 3: 10^{17}, 4: 10^{18}) [320].

Bakay and Kiryukhin [321] proposed the model of evolution of precipitates under irradiation of supersaturated solid solutions. This model takes into account the following, the dissolution of precipitates by means of their overlap with cascades of displacements, the nucleation and growth of precipitates aside from the initial precipitates, and the process of coalescence. Their estimations provide the value of precipitate nucleation time ≈ 10 s, the time of breakdown in the cascade $\approx 10^3$ s and the time of coalescence process progress $\approx 10^5$ s. The following expression was received for the growth rate of large precipitates:

$$\frac{dr}{dt} = \frac{D}{r}\left(\Delta - \frac{pl_c l_p}{D} - \frac{\alpha}{r}\right).$$ (7.30)

Here r is a radius of precipitate, D is the diffusion coefficient of solute atom, $\Delta = n_0 - n_{th} - n_p$, n_0 is the total average concentration of solute according to volume, n_p is the average concentration of solute according to volume, included in the precipitates, p is the probability of fracture of precipitate with layer thickness l_p, l_c is the radius of cascade,

$$\alpha = n_{th} \frac{\sigma\Omega}{kT} \cdot$$

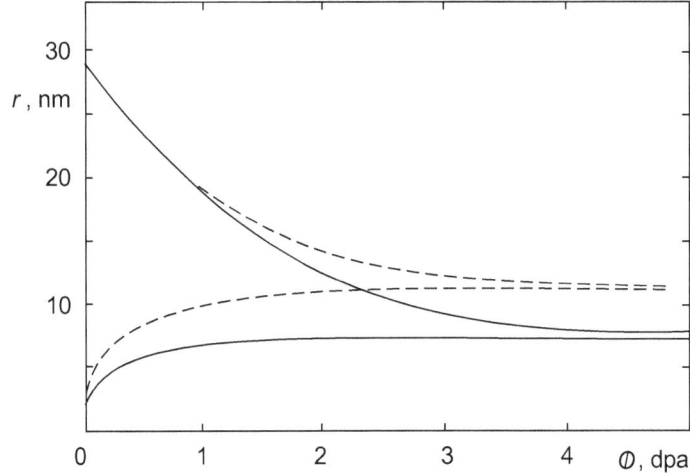

Fig. (7.9). The change in precipitate radius (r) with dose. Continuous curves for $K = 10^{-2}$ and dashed curves for $K = 10^{-6}$ dpa s^{-1} [320].

In case of cascade dissolution of precipitates, i.e. if $pl_p l_c / D \gg \alpha/r$ the radius of precipitate under irradiation approximates to the following quasi-equilibrium value ($2R$ is the distance between precipitates)

$$r^{irr} = R\sqrt[3]{n_0 - \frac{pl_c l_p}{D}} \cdot \tag{7.31}$$

It is clear that if the intensities of radiation are very high $r^{irr} \to 0$, then all large precipitates will be dissolved. With the temperature decrease the diffusion coefficient decrease, that results in the acceleration of dissolution of large precipitates.

Since the effective supersaturation of solid solution in the presence of cascades is

$$\Delta' = \Delta - \frac{Kl_c l_p}{D},$$

the critical radius of precipitate under irradiation

$$r_{kr}^{rad} = \frac{\alpha}{\Delta'} \tag{7.32}$$

will be higher than without irradiation.

For new small precipitate aside from the displacement cascades, the growth rate is defined by the following:

$$\frac{dr}{dt} = \frac{D}{r}\left(\Delta - \frac{\alpha}{r}\right) \cdot \tag{7.33}$$

The high supersaturation of solid solution as a consequence of radiation-induced dissolution of large precipitates results in the second maximum in distribution function considering in the area of small sizes. Such transfer of solute from the large precipitates to small ones is in fact a ***radiation inverse process of coalescence.***

The latter model describes the experimental results of Bakay *et al.* [322] quite well. Bakay and colleagues presented the results of exploration connected with the impact of irradiation by ions of chrome, on the distribution of precipitate sizes for the γ'-phase (Ni$_3$(Ti,Al)) in the PE-16 alloy. After annealing of samples at the temperature of 1023 K, the precipitates of γ'- phase generated with a wide spectrum of sizes and with the maximum diameter of 45 nm. The following irradiation at the temperature of 850 K practically kept the microstructure of samples unchanged. After irradiation at the temperature of 690 K, two maximums appear in the size distribution at 35 nm and at 6 nm. The irradiation at the temperature of 590 K with the dose of 37 dpa results in full dissolution of initial large precipitates.

Abyzov *et al.* [323] considered that the appearance of size limit of a precipitate may be caused not only by knocking out individual atoms from the surface layers of precipitate or diluting action of cascades, but also by means of diffusion dissolution of precipitates under irradiation.

7.3.5. Decomposition of Spinodal Type

Krishan and Abromeit [324] considered the radiation-stimulated decomposition of solid solutions of spinodal type. The model proposed by authors is based on kinetic equations of local accumulation of any component in the AB alloy, defined by the movement of interstitial atoms of both components. The local enrichment depends on the radiation-induced by locally rates of generation of interstitial A and B atoms, their annihilation rate with vacancies, interrelation of space distribution of local concentrations of A and B components and their local interstitial states. The possibility of formation of component-radiation defect complexes or any traps for defects is not analyzed in the model. According to their model the wavelength of concentration fluctuations is determined by the following expression:

$$\lambda \approx 2\pi \left(\frac{1}{Kn_v|F|} \right)^{1/2}, \tag{7.34}$$

where $K = 4\pi r_r/\Omega$, r_r is the recombination radius, n_v is the averaged vacancy concentration, F is the parameter connected with the defect formation rate.

Wagner *et al.* [325] performed the analysis of spindle decomposition, irradiated by the electrons of Ni-Cu alloy, using the kinetics of short-range clustering in good accordance with Cook's theory.

7.3.6. Radiation Effect on the Kinetics of Impurity Aggregation in AHC

We investigated the impact of isotopic β-radiation on the kinetics of aggregation of tin, copper and gallium in some AHCs at the room temperature [326]. Kinetic curves were processed using the integral equation of aggregation kinetics of the second order. The rate constants without irradiation (k) and irradiated (k^{irr}) and equilibrium concentrations without irradiation (n) and irradiated (n^{irr}) were determined by matching their values with the best fitted calculated curves with experimental data on the Gauss criterion. Table **7.2** presents the relationships n^{irr}/n and k^{irr}/k for the researched samples. This data suggest that the irradiation causes changing of equilibrium concentration and the constant of aggregation rate. The investigated solid solutions may be divided in three groups considering their radiation impact. The decrease of solubility limit and the increase of aggregation rate is typical for AHC, doped by tin. The growth of solubility limit and the decrease of aggregation rate is observed for AHC with the copper. In case of cesium bromide doped gallium, the irradiation decreases the solubility limit and the aggregation rate. In view of the impact of binding energy of the impurity-radiation defect complex on the solubility limit [316], we may assume that the binding energy of radiation defects with gallium is small, with tin is bigger and with copper is yet bigger. The radiation change of aggregation rate constant is connected with the change of diffusion coefficient. The transfer of Sn^{2+} ions in the AHC is realized by using the vacation mechanism. Therefore, the increase of rate constant of aggregation is undoubtedly caused by the radiation formation of cationic vacancies. The radius of Cu$^+$ and Ga$^+$ ions is small (≈ 0.1 nm). It is close to the radius Na$^+$ (0.097 nm), less than the radius K$^+$ (0.133 nm) and far less than the radius Cs$^+$ (0.167 nm). It is necessary that the ion of copper in the potassium chloride occupies the off-center position in the lattice site, and the gallium ion is located in the cesium bromide in the interstitial. Therefore, the movement of copper and gallium ions is partly realized by using the interstitial mechanism. Apparently, the formation of radiation defects in the AHCs results in decreasing of the diffusion coefficient for interstitial ions.

Table 7.2.   The Relative Equilibrium Concentrations and Constants of Aggregation Rate During Irradiation for Certain Doped Alkali Halides [326]

System	T, K	n^{irr}/n	k^{irr}/k
KCl-Sn	293	0.84	1.76
KBr-Sn	393	1.00	1.48
KBr-Sn	413	0.88	4.19
KBr-Sn	423	0.83	1.19
KI-Sn	373	0.56	1.13
NaCl-Cu	403	1.16	1.00
NaCl-Cu	423	1.05	0.72
KCl-Cu	393	1.07	0.74
KCl-Cu	413	1.24	0.73
KCl-Cu	433	1.22	0.64
KCl-Cu	453	1.08	0.60
CsBr-Ga	413	0.60	0.39
CsBr-Ga	453	0.51	0.35

CHAPTER 8

Radiation-Stimulated Segregation

Abstract: The experimental data and models of radiation-induced segregation are presented. The mechanism and efficiency of the radiation-induced segregation depend on several factors: the irradiation temperature, the concentration and size of atoms of impurity, the presence of additional impurity, *etc*. The majority of models which describe the radiation-induced segregation, include a system of kinetic equations for the various kinds of point defects and alloy atoms. The models differ from each other in a set of the types of defects and calculation parameters.

Keywords: Radiation, metal, alloys, alkali halide, solid solution, segregation, segregation models, inverse Kirkendall effect.

The radiation-stimulated segregation is one of the most important phenomena that occur in the solid solutions under irradiation. Normally, the radiation-stimulated segregation results in degradation of physical, chemical and mechanical properties of solid materials. On the other hand, in some cases segregation suppresses swelling. The segregation is an accumulation of solute near the linear, plane and volume defects of crystal lattice. The radiation-stimulated segregation means the enrichment or depletion of solid solution areas near the sinks of pointed defects (surface of sample, surface of micro-modules, grain boundaries, dislocations, dislocation loops, and caverns) by the solute under irradiation. At the same time, there is a reverse process, i.e. the bulk of the sample (defect-free regions of lattice) is depleted or enriched by impurities. The driving force of radiation-stimulated segregation is the supersaturation by radiation point defects. The necessary condition for realization of segregation is a rather high mobility of solute atoms. The movement of solid solution components into or out of sinks may be realized in the form of separate atoms, in the form of complexes solute-radiation defect and by dragging the solute atoms by flows of point defects. The radiation-stimulated segregation may be realized when the concentration of solute is higher or lower than the solubility limit. For example, the segregation takes place during the irradiation of reactor neutrons ($\Phi = 2.4 \ 10^{23} \ \text{m}^{-2}$) at temperature of 325 K in the Fe-Ni alloy, where the nickel content is lower than the solubility limit [327].

The radiation-stimulated segregation was apparently described for the first time by Anthony [328]. The development of radiation segregation phenomenon may be pictorially shown by experiments with various alloys. The direction and effectiveness of segregation depends on the relative size of alloy atoms (more or less than the size of matrix atoms), on the conditions of irradiation (kind and power of radiation, temperature), on the solute concentration (diluted or concentrated solutions) and on other factors.

8.1. EXPERIMENTAL DATA

8.1.1. Temperature Dependence

Geits and Natsvlishvili [329] found that the irradiation of doped LiF by neutrons with fluence of $10^{20} \ \text{m}^{-2}$ ($\phi = 1.2 \ 10^{16} \ \text{m}^{-2} \ \text{s}^{-1}$) at 310 K causes segregation of solute on dislocations accompanied with the formation of clusters with the size ≈ 0.1 micron. Irradiation of the same at 110 K leads only to the formation of dislocation loops and associated radiation defects. Under irradiation by small doses at 110 K and 310 K only pointed defects are generated. In general, the radiation-stimulated segregation in the solid solutions on the basis of alkaline halides is practically not researched. To realize this phenomenon high mobility and concentration of impurity atoms and pointed defects are necessary. However, in AHCs in these conditions, the processes of aggregation of pointed defects take place with high intensity, resulting in the radiolysis of matrix.. We may expect the radiation-stimulated segregation only in the most stable alkaline halides (for example, LiF).

Turos and Meyer [330] studied the segregation of gold, implanted in iron under irradiation with the He$^+$ ions (2 MeV). It was found that at relatively low temperatures (350-500 °C) the solute-vacancy complexes form only the gold precipitates (due to the low mobility of complexes). At intermediate temperatures (550-650 °C) the migration of complexes gold-vacancy causes departure of solute to the surface, i.e. the segregation takes place. At the temperatures above 650 °C we observe the inverse process when the Au diffuses backwards from the surface to the bulk of sample.

According to data from Clausing *et al.* [331] irradiation at the temperature of 410 °C by neutrons with the dose of Φ = 13 dpa causes strong segregation of Ni, Cr, Si and P on the grain boundaries in the martensitic steel. At the same time, when the temperature is 520 °C and 565 °C, the segregation is not detected.

Zhang *et al.* [332] investigated the impact of irradiation temperature (1 MeV electrons) on the segregation of zinc in the Al-0.35at.%Zn alloy. It appeared that at room temperatures the segregation rate increases with temperature (energy of activation of 0.15 eV). Within intermediate temperatures the rate of zinc segregation doesn't depend on temperature, and the segregation rate decreases at high temperatures. It is assumed that segregation takes place due to the movement of zinc-interstitial complexes towards sinks. At low temperatures, the effectiveness of segregation is determined by recombination of pointed defects. When the temperatures are high, the process of defect movement towards sinks is dominating, and if the temperatures are yet higher, the dominant factor is the thermal diffusion of vacancies. The same authors [333] in the same irradiation conditions investigated the Al-Zn alloy with a higher content of solute. It appeared that when the content of zinc is 4.5 at. %, irradiation by electrons at the temperatures of 80-130 °C results in the formation of small metastable and spherical Guinier-Preston zones and plate shaped precipitates on (111) planes throughout the volume.

8.1.2. Dependence on Size of Alloy Atoms

Nakata and Masaoka [334] found that under irradiation of austenitic steels by the 1MeV electrons (ϕ = 4.8 10^{23} m^{-2} s^{-1}, Φ = 5 10^{27} m^{-2}), undersized solute (Ni, Si or P) segregate towards the grain boundary, and an oversized solute (Cr or Mo) segregate away from the boundary. The segregation increases with irradiated temperature from room temperature to 773 K, except for Si. It was also found that the grain boundary frequently migrates during irradiation at temperatures 673-873 K.

Agarwal *et al.* [335] explored segregation of chrome in the V-15wt.%Cr and found that irradiation by the V$^+$ ions with an energy of 3.5 MeV with the doses of 15 and 60 dpa results in detectible chrome output towards the surface of sample. The enrichment by solute down to the depths of 20 nm and the depletion at the depth of 20-100 nm are observed.

Wang *et al.* [336] presented the results of their investigation of segregation in the titanium alloy at 650 °C under irradiation by the 2.1 MeV Ar$^+$ ions (dose rate close to surface $\approx$ 8 10^{-4} dpa s^{-1}). The radiation-stimulated enrichment of the near-surface layer was detected for three elements V, Mo and Al (undersize solute). The segregation from bulk towards the surface is strongly expressed for V, for Mo it is expressed weaker, and for Al it is insignificant. The present segregation of undersize solutes toward sinks is in accord with the size factor. For undersize solutes, the segregation mechanism is likely to be the migration of interstitial-solute complexes toward point-defect sinks.

Allen *et al.* [337] studied the radiation-stimulated segregation of undersize and oversize solutes in the stainless steel. The irradiation by the 3.2 MeV protons at 400 °C with the dose of 1 dpa caused typical enrichment of space near the sinks (voids) by the undersize solute (Ni) and depletion by the oversize solute (Cr). The distribution of solutes according to the distance from a void is presented in Fig. (**8.1**).

It is interesting to note that irradiation of stainless steel Fe–12Cr–15Mn by the 1.25 MeV electrons (ϕ = 5 10^{-2} dpa s^{-1}, Φ = 5.4 dpa) at the temperatures of 300-600 °C causes segregation of manganese (undersize) at the grain boundaries [57], but it doesn't result in depletion of these boundaries by chrome (oversize).

Ashworth *et al.* [338] exploring radiation-stimulated segregation in the Fe-20Cr-25Ni-Nb alloy under irradiation by electrons, observed the typical enrichment of grain boundaries by Ni and Si as well as the depletion by Cr, Fe and Mo. In addition, they detected the radiation-stimulated movement of grain borders. This phenomenon was observed even with the doses of 1.8 dpa.

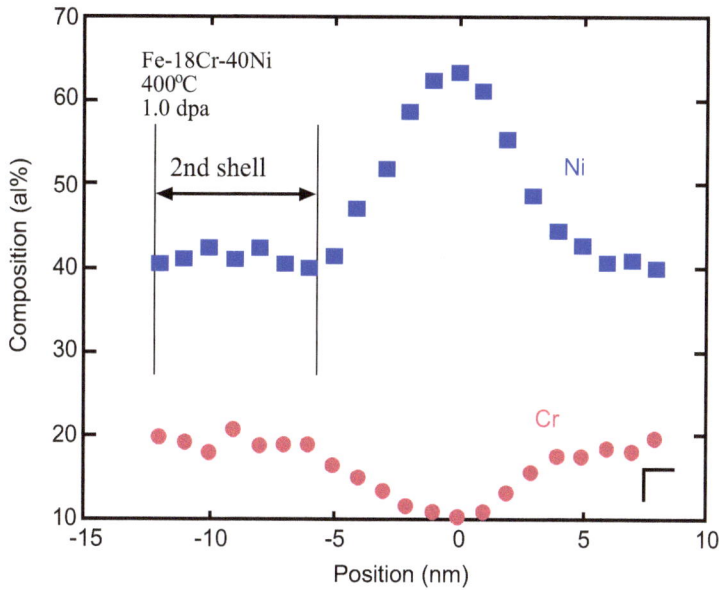

Fig. (8.1). Radiation-induced segregation in an irradiated with 3.2 MeV protons Fe-Cr-Ni alloy [337].

Hautojarvi *et al.* [339] observed the segregation of carbon (undersize) on the vacancy clusters during annealing at the temperature of 350 K after irradiation at the temperature of 77 K by fast neutrons ($\phi = 3 \cdot 10^{15}$ m^{-2} s^{-1}, $\Phi = 4 \cdot 10^{20}$ m^{-2}) in the iron with carbon solute ($n_C = 5 \cdot 10^{-5}$ and $7.5 \cdot 10^{-4}$).

Kornblit and Ignatiev [340] reviewed the literature and made a conclusion that the radiation-stimulated segregation takes place only if the volume difference (size factor, δ_V) is less than zero. If r and r_M are the radiuses of solute and matrix atoms, respectively, then

$$\delta_V = \left(\frac{r}{r_M}\right)^3 - 1 > 0,$$

the segregation is impossible. Table **8.1** presents the values of size factor for some systems. By claiming that for an oversize solute the segregation is impossible, Kornblit and Ignatiev meant apparently, the classical segregation is the output of alloy to the boundaries of space defects of solid solution.

Okamoto and Rehn [341] reviewed the experimental data on segregation in irradiated materials and showed that in most cases the concentration of undersize atoms rises in regions close to sinks and falls elsewhere, while the concentration of oversize atoms falls in regions close to sinks and rises elsewhere.

Table 8.1. Values of Size Factor for Some Alloys [341, 342]

Solvent	Pd	Pd	Cu	Ni	Ni	Ni	Ni	Ti	Ti	Ti
Solute	Cu	Fe	Ag	Al	Cr	Al	Cr	V	Mo	Al
δ_V	-0.20	-0.27	+0.44	+0.52	+0.01	+0.52	+0.01	-0.15	-0.21	-0.20

Wang *et al.* [343] studied the radiation-induced segregation in the titan alloys and found that all major features of radiation-induced segregation in the alloys with hcp structure are qualitatively similar to those reported for alloys with the fcc structure. Undersize solutes due to the point defect fluxes cause enrichment in the near-surface region. Oversize solutes cause depletion in the near-surface region. However, effectiveness of segregation cannot correlate quantitatively with the size factor. For example, molybdenum and aluminum with the size factor of -0.21 and -0.20, correspondently, segregate weaker than vanadium, for which $\delta_V = -0.15$. Or, tin with $\delta_V = +0.14$ doesn't demonstrate any significant redistribution of solute under irradiation. Therefore, it can be concluded that for hcp and for fcc alloys, the volume of size factor can be used as a qualitative guide, but the quantitative prediction based on size effect alone is not possible.

8.1.3. Influence of Small Impurity Addition on the Segregation of Main Alloy Component

The radiation-stimulated segregation of main alloy component is sometimes influenced significantly by a small addition of any impurity. Hackett *et al.* [342] investigated the radiation-stimulated segregation of chrome in the stainless steel under irradiation by the 3.2 MeV protons ($\phi \approx 10^{-5}$ dpa s^{-1}). As it was expected, the irradiation results in depletion by chrome near the sinks of point defects. Addition of 0.038 wt. % Zr significantly suppresses the depletion by Cr near the grain boundaries. At the same time, the suppression of segregation takes place only up to Φ = 3 dpa at 400 °C and up to Φ = 7 dpa at 500 °C. If the doses are larger than the mentioned ones the effect of suppression disappears. The reason of segregation suppression is a slowdown of chrome migration due to the formation of Zr-vacancy complexes. The addition of hafnium also suppresses the radiation-stimulated segregation, but the effectiveness of such suppression is significantly smaller than for zirconium. It is proposed that the binding energy of Zr-vacancy is larger than the binding energy of Hf-vacancy.

The peculiarities of hafnium to suppress radiation-stimulated segregation of phosphorus were noted by Lu *et al.* [344]. The irradiation of E911 alloy by the 250 keV Ni$^+$ ions at the temperature of 300 °C causes the segregation of phosphorus at the grain boundaries with the value of 1.66 wt. %. The segregation of phosphorus in the E911 alloy with the 1 % hafnium in similar irradiation conditions amounted to 0.27 wt. %.

Kato *et al.* [345] found that when the stainless steel 316 is exposed to radiation by the 1MeV electrons ($\phi = 4.2 \cdot 10^{23}$ m^{-2} s^{-1}, $\Phi = 10.8$ dpa) at the temperature range of 623-773 K, intensive enrichment of grain boundaries by nickel and depletion by chrome take place. Addition of oversize Zr and Hf significantly suppresses the segregation of Ni and Cr. The observed effect may be explained by the capture of point defects by the oversize solute, as it causes their intensive recombination, and decrease the fluxes of point defects toward the sinks.

8.1.4. Other Features of Radiation-Stimulated Segregation

Hayashi *et al.* [1] received interesting results when they explored the impact of irradiation on the amorphous alloy $Fe_{100-x}B_x$ ($x \leq 25$). The irradiation by the 40 keV He$^+$ ions at the temperature below 160 °C results in the formation of crystal phase α-Fe at the surface of amorphous alloys $Fe_{80}B_{20}$ and $Fe_{85}B_{15}$. It was also determined that the quantity of segregated α-Fe is decreasing from the surface towards the depth of the irradiated samples. At the same time, the $Fe_{75}B_{25}$ alloy remained stable against the radiation impact. Apparently, irradiation in this experiment induces the decomposition of solid solution on phase α-Fe and Fe_3B. On the other hand, the enrichment of near-surface layers with iron under irradiation is typical for the radiation-induced segregation. In majority of cases a solute segregation takes place, but in this case we observe the segregation of a solvent.

The decomposition of solid solution during the radiation-stimulated segregation was observed by Yoshida *et al.* [346] under irradiation by the protons (2.3 MeV, $\phi = 2.2 \cdot 10^{18}$ m^{-2} s^{-1}, $\Phi = 1.3 \cdot 10^{23}$ m^{-2}) of Au-1.0at.%Fe alloy at the temperature of 220 °C. Using the method of Mössbauer spectroscopy, it was detected that half of the iron atoms participating in segregation was represented by small complexes of iron.

In some cases we observe the dependence of segregation from the concentration of solute. For example, in Pd-Fe alloy the radiation-induced precipitation of Pd_3Fe on the sinks was observed only at 18 at. % concentration of iron between 110 to 500 °C, irradiated with 400 keV protons to a dose of 0.9 dpa [347]. Irradiation-induced precipitation was not observed in the Pd-2.8at.% and 12at.%Fe alloys proton irradiation to the same dose, nor to a higher dose of 1.5 dpa. The segregation also was not observed in the 2 and 8 at. % Fe alloys irradiated at 600 and 700 °C by 3 MeV Ni$^+$ ions.

It was found that the irradiation causes significant segregation of non-metals: phosphorus, sulfur, oxygen and other elements. Brimhall *et al.* [348] established that the irradiation by the 5 MeV Ni^{2+} ions even with small doses of 0.01-0.8 dpa causes the segregation of phosphorus towards the surface of samples in steel at the temperatures of 675-875 K.

Arbuzov *et al.* [349] observed the output of sulfur to the surface in nickel under irradiation by the 5 MeV electrons with fluence of 10^{23} m^{-2}. It turned out that the irradiation by electrons at 350 °C up to the dose of 10^{23} m^{-2} results in 10-fold depletion of near-surface layers (> 0.4 micron) by sulfur in relation to the original content.

The results of experimental researches demonstrate that the effectiveness of segregation with close absorbed doses is higher for irradiation by particles with a smaller stopping power. Thus, Erck and Rehn [350] found that under irradiation by the 1.8 MeV He$^+$ ions ($\phi = 10^{-4}$ dpa s^{-1}) the value of segregation in the Mo-Re alloy is nearly proportional to the absorbed dose ($\Phi = 0.05\text{-}1$ dpa), and it is significantly higher than in case of irradiation by the 1.8 MeV Ne$^+$ ions ($\phi = 6 \ 10^{-4}$ dpa s^{-1}).

The irradiation of low alloyed steel by fast neutrons ($\phi = 10^{17}$ m^{-2} s^{-1}, $\Phi = 10^{24}$ m^{-2}) results in decrease of temperature for the beginning of sulfur segregation from 400 °C to 300 °C [351].

The phenomenon which is similar to radiation-induced segregation was observed under irradiation of the Ni-12.6at.%Si alloy by electrons [352]. It was determined that the Ni$_3$Si precipitates placed near the surface are growing, but the precipitates at the bulk of sample are dissolving. The dissolution energy equals to 0.5 eV, and it is 2 times lower than the energy of vacancy migration.

8.2. MODELS OF RADIATION-STIMULATED SEGREGATION

At the present moment, the researchers use two basic models while calculating the radiation-stimulated segregation, in the basis of which lays the following: (i) Fluxes of solute-radiation point defect complexes and (ii) Inverse Kirkendall effect. In the majority of cases the researchers observe the models of radiation segregation in the binary dilute alloys. Here the term 'dilute' indicates that the solute atoms in the solid solution are separated by more than 2 atomic spacings. Usually, this condition is carried out at the concentration of an impurity less than 1 at. %.

8.2.1. Radiation-Accelerated Segregation

Yin *et al.* [353] presented the results of modeling of carbon segregation in the ferritic steel under irradiation by neutrons. It is proposed that irradiation only accelerates the process of Fe$_3$C precipitation at the grain boundaries without an impact on the driving force of solid solution decomposition. The effect of irradiation on the origin and growth of precipitates occurs through the radiation acceleration of diffusion coefficient. To calculate the solute diffusion coefficient under irradiation the following formula was used:

$$D_a^{irr} = D_a \frac{n_v^{th} + n_v^{irr}}{n_v^{th}},\tag{8.1}$$

where D_a is the solute diffusion coefficient without irradiation, n_v^{th} is the concentration of vacancies without irradiation, n_v^{irr} is the stationary concentration of vacancies, generated by irradiation. In contradistinction from the majority of other models, the calculation of n_v^{irr} is performed without the kinetics rate equations. The concentration of vacancies was calculated as follows:

$$n_v^{irr} = \frac{KF}{D_v k_v^2},\tag{8.2}$$

$$F = \frac{2}{\eta}\left[(1-\eta)^{\frac{1}{2}} - 1\right],\tag{8.3}$$

$$\eta = \frac{4\alpha K}{k_v^2 k_i^2 D_v D_i},\tag{8.4}$$

$$k_v^2 = \rho^{\frac{1}{2}}\left(\frac{3}{R} + \rho^{\frac{1}{2}}\right),\tag{8.5}$$

$$k_i^2 = (z_i\rho)^{\frac{1}{2}} \left[\frac{3}{R} + (z_i\rho)^{\frac{1}{2}} \right],$$

(8.6)

$$\alpha = \frac{21D_i}{\lambda^2}.$$

(8.7)

Here K is the rate of radiation formation vacancies, D_v and D_i is the thermal diffusion coefficient of vacancies and interstitial atoms, respectively, k_v^2 and k_i^2 is the sink efficiency for vacancies and interstitials, respectively, ρ is the density of dislocations, R is the radius of grain, z_i is the advantage factor in interaction of interstitials with the dislocations in relation to the vacancies ($z_i = 1.1$), λ is the jump distance of the self-interstitials. The results of calculations received on the basis of the proposed model are in good agreement with the experiment data. It is also determined that high contents of nickel and copper impacts on the segregation of Fe_3C at the grain boundaries. Nickel suppresses and the copper accelerates this process. This is explained by the fact that nickel increases the solubility of carbon in iron and copper on the contrary.

8.2.2. Simplified Model

Pechenkin and Epov [354] theoretically examined the radiation-stimulated segregation in the substitutional alloys at the sinks of point defects. Their model doesn't consider any interaction of solute with point defects accompanied by the formation of solute-defect complexes. They used a system of equations for the fluxes of impurities and point defects, proposed by Lam *et al.* [355], but without the equations and terms, describing the formation and development of impurity-defect complexes. The calculations were executed for three types of sinks: surface of foil, voids, and dislocations. If the partial diffusion coefficients for the profile of component A are not differ significantly near the border of sink, the following formula is given:

$$n_A(r) = \frac{Cn_v^{irr}(r)}{1 + Cn_v^{irr}(r)}.$$

(8.8)

Here C is the normalizing factor, $n_v^{irr}(r)$ is the stationary concentration of vacancies under irradiation, depending on the distance r from the sink. From Eq. (8.8) we can make a conclusion that with the increase of distance from the vacancy sink the concentration of A component increases, since the concentration of vacancies is growing. The calculated concentration profiles have an appearance which is indicative for the radiation-induced segregation with the depletion of near-surface layers by A component.

8.2.3. Diluted Solid Solutions

Model of Johnson and Lam

Johnson and Lam were the first who developed the model describing the solute segregation in irradiated substitutional alloys with an fcc lattice [356]. This model was based on kinetic equations for the fluxes of solute atoms, self-interstitial atoms and vacancies. The calculation used the parameters which were appropriate for Zn in Ag. The most intensive segregation observed was in the temperature range from 0.2 to 0.6 T_m (T_m is melting temperature). Maximum segregation is observed around 0.45 T_m for a defect-production rate of 10^{-3} dpa s^{-1}. The migration of solute towards the foil surface using the interstitial mechanism is the basic one in the process of segregation. With an interstitial-solute binding energy of 0.2 eV, solute segregation is found to decrease as the vacancy-solute binding energy increases. The initial solute concentration doesn't affect significantly the quantity of $\Delta n/n_0$ near the surface of solid solution.

Later, Johnson and Lam upgraded their model. Their new model [357] considered additionally the transfer of solute using solute-point defect complexes. The model also takes into account the interaction of solute with the fluxes of point defects (solute entrainment by the fluxes of defects). The calculations were performed on the basis of system of kinetic differential-integral equations with the variation of defect formation rate, binding energy of the complex solute-pointed defect (E_{ai}, E_{av}), sink types (plane surface, voids, dislocations), considering the direct and reverse reactions. The change of solute concentration from the time is described by the following equation:

$$\frac{\partial n_a}{\partial t} = \nabla(\sigma_i D_i n_a \nabla n_i - \sigma_v D_v n_a \nabla n_v) - k_2 n_a n_i + k_2' n_{ai1} - k_3 n_a n_i + k_3' n_{ai2} - k_4 n_a n_v - k_4' n_{av} + k_5 n_v n_{ai1} + k_6 n_v n_{ai2}.$$

(8.9)

Here σ_i and σ_v is the force of solute interaction with the fluxes of interstitial atoms and vacancies, respectively, n_i, n_v, n_{ai1}, n_{ai2}, n_{av}, is concentration of interstitial atoms, vacancies, solute-interstitial atom complexes of two types *1* (mobile) and *2* (immobile), solute-vacancy complexes, k_j and k_j' are the rate constants of direct and reverse reactions, respectively. The rate equations for the concentrations of point defects have the appearance which is similar to equation (8.9). The calculation was performed using parameters corresponding to the segregation of Zn in Ag. The most important parameters had the following values: the migration energy of interstitial and vacancy amounted to 0.10 and 0.84 eV, the binding energy of complexes was $E_{ai} = 0.20$ eV and $E_{av} = 0.05$ eV, the initial solute concentration was 10^{-3}. Two characteristics of segregation were detected: the enrichment of sink surface and the depletion by solute at the half of average distance between the sinks. For example, this distance was accepted as 50 nm for the foil.

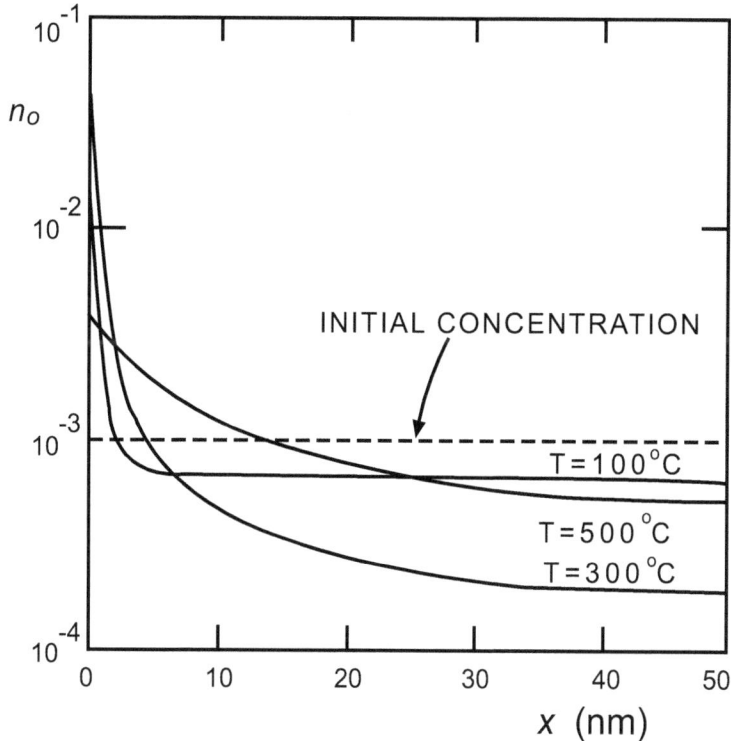

Fig. (8.2). Steady-state concentration of solute atoms as a function of distance from a foil surface [356].

The performed calculations demonstrated (see Figs. **(8.2)** and **(8.3)**) that the segregation takes place only in the specific temperature range, depending on the dose rate. The degree of enrichment may reach one hundred, and the degree of depletion may reach 150. With increase dose rate the temperature of maximum segregation (T_s) is shifting to the area of high temperatures. At the same time, the degree of enrichment and depletion for the smaller dose rate is higher than for the larger dose. The effect of temperature on segregation at a given irradiation rate can by explained as follows. At the low temperatures $(T < 0.2\ T_s)$, the vacancies are immobile and do not go to the sinks. As a consequence, their concentration increase and the interstitial atoms recombine with them and do not reach the sinks. The absence of fluxes of interstitials excludes segregation. At a higher temperature $(T > 0.6\ T_s)$, thermal vacancy concentration becomes comparable to radiation-induced vacancy concentration and the recombination of interstitials with vacancies becomes defining. Both the interstitial and solute fluxes to sinks are absent. The effectiveness of segregation significantly depends on the binding energy of solute-interstitial atom complex and has a maximum value when $E_{ai} = 1.25$ eV.

In the same conditions, the enrichment for voids may be significantly larger than the enrichment for foils. In the case of voids, the segregation depends on the void radius (r_v). The maximum enrichment is achieved when $r_v = 20$ nm. The depletion doesn't have a maximum and it transfers to saturate when r_v is growing.

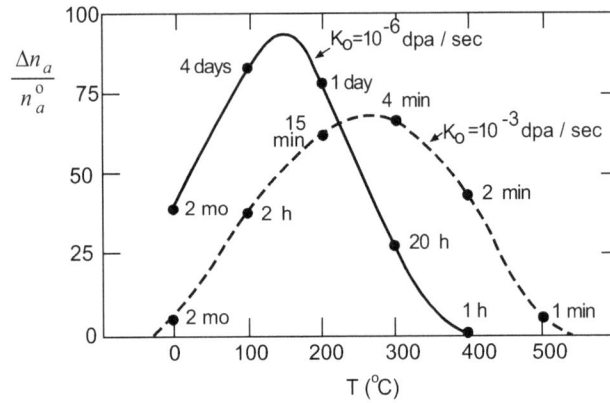

Fig. (8.3). Temperature dependence of solute segregation at the two defect-production rate [357].

Model of Murphy

Murphy [358] proposed a model of radiation segregation for the diluted binary alloys, which is based on the diffusion kinetic theory and according to his opinion provides more accurate equations intended for fluxes of solute and defects than Johnson-Lam's model. The model used here includes isolated solute atoms, vacancies and interstitials, and solute-defect complexes, which consist of a vacancy or an interstitial on a site neighboring a solute atom. It is assumed that the presence of the complexes does not affect the efficiency of recombination of point defects, and the interaction of fluxes of different defect types is negligible.

In the irradiated alloy, the full concentration of solute atoms, vacancies and interstitials is determined by the following equations:

$$\frac{\partial N_a}{\partial t} = -\nabla J_t, \tag{8.10}$$

$$\frac{\partial N_v}{\partial t} = K - \alpha N_i N_v - L_v - \nabla J_v, \tag{8.11}$$

$$\frac{\partial N_i}{\partial t} = K - \alpha N_i N_v - L_i - \nabla J_i, \tag{8.12}$$

where J_a, J_v and J_i are the fluxes of solute, vacancies and interstitials, respectively, K is the production rate of the point defects, α is the recombination coefficient, L_v and L_i are the loss rates of vacancies and interstitials to the sinks, respectively (*e.g.* to dislocations and grain boundaries). Neglecting the direct production of solute-defect complexes, the fluxes of species are given by equations:

$$J_v = -D_v \nabla n_v - D_{av}^v n_v \nabla n_v - D_{va}^v n_v \nabla n_t, \tag{8.13}$$

$$J_i = -D_i \nabla n_i - D_{ai}^i n_i \nabla n_i - D_{ia}^i n_i \nabla n_t, \tag{8.14}$$

$$J_a = -(D_{va}^t \nabla n_v + D_{ia}^t n_i) \nabla n_a - D_{av}^t n_a \nabla n_v - D_{ai}^t n_a \nabla n_i, \tag{8.15}$$

where n_v, n_i and n_t are the concentrations of free vacancies, interstitials and solute atoms respectively, the coefficients D_{av}^v, *etc.* are functions of the various jump frequencies for species. The superscripts are used to label coefficients for the flux equations for species. The subscripts are used to indicate that the coefficients are multiplied by species concentrations (first subscript), or by species gradients (second subscript).

The calculation of radiation-stimulated segregation was carried out by a numerical method. The whole volume was divided into separate cells (with the size up to 5 nm) with free surfaces. Contrary to the Johnson-Lam's calculations [357], where it was proposed that the free surface is an ideal sink, i.e. the concentrations of vacancies and interstitials near the sink surface equal to the thermally equilibrium concentrations, here the gradient of concentrations of pointed defects (and consequently, fluxes) near the surface depends on the irradiation rate. The assumption of rate-limitation on the absorption of vacancies and interstitials is physically more realistic. The calculations are carried out for case of manganese segregation in nickel during nickel ion irradiation. In this case the impurity migrates away from the free surface into the bulk. In the calculations it is assumed that the vacancy migration energy in nickel is 1.28 eV since this value gave results for segregation which agreed more closely with the Hobbs-Marwick's experimental results [359]. The experimental and calculated concentration profiles of manganese in nickel are presented in Figs. (**8.4**) and (**8.5**). It is obvious that the theoretical and experimental data are in line with each other.

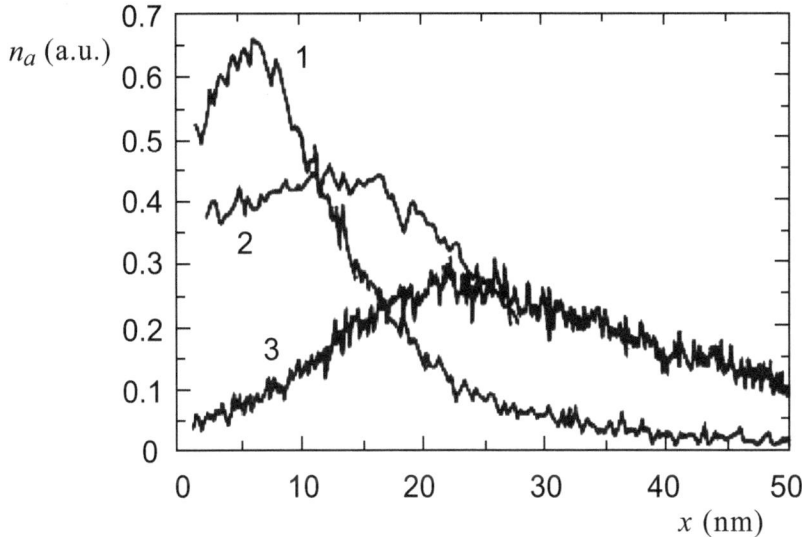

Fig. (8.4). Experimental results for segregation of Mn in Ni during ion-bombardment from Hobbs and Marwick [359]. 1: As implanted; 2: 7.2×10^{19} m^{-2}; 3: 7.2×10^{19} m^{-2}; x is distance from surface.

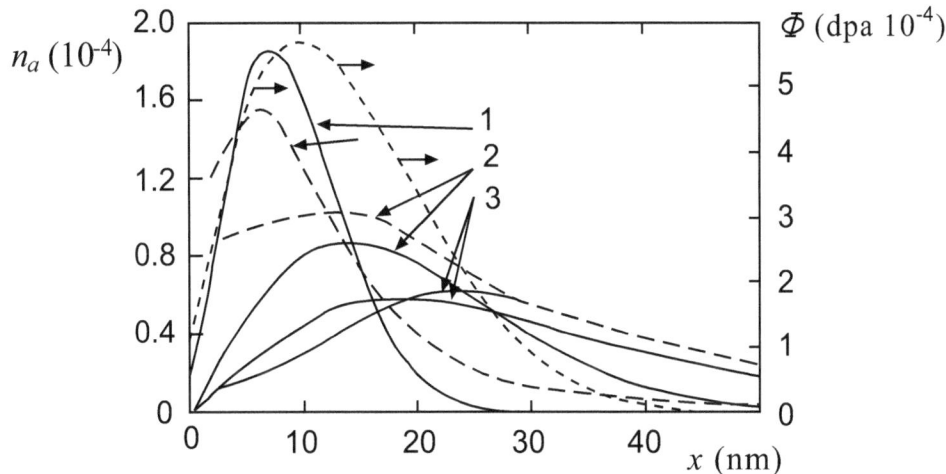

Fig. (8.5). Result of theoretical calculation for the segregation of Mn in Ni [358]. 1: Initial; 2: 7.2×10^{19} m^{-2}; 3: 7.2×10^{19} m^{-2}; x is the distance from surface. Dashed curves: Experimental results of Hobbs and Marwick [359]. Short dashed curve: Calculated curve for the displacement damage.

Model of Bartels, Dworschak and Weigert

Bartels *et al.* [360] investigated the kinetics of radiation-stimulated segregation in the dilute alloys Ni-Si and Ni-Ge under electron irradiation (3 MeV, $\phi = 5.8 \times 10^{19}$ m^{-2} s^{-1}) using the measurement technique of residual electrical resistivity. It was determined that at the temperature of 535 K 2-10 Frenkel pairs must be produced to segregate one silicon atom in the Ni-Si alloy and at least 260 Frenkel pairs must be produced to segregate one germanium atom in the Ni-Ge alloy. To describe the received results the researchers used the set of rate equations for concentration of the interstitials, vacancies and solute-self-interstitial complexes:

$$\frac{\partial n_i}{\partial t} = K - k_{iv} n_i n_v - k_{ai} n_a n_i - k_{is} n_i n_s + v_c n_c; \tag{8.16}$$

$$\frac{\partial n_v}{\partial t} = K - k_{iv} n_i n_v - k_{cv} n_c n_v - k_{vs} n_v n_s; \tag{8.17}$$

$$\frac{\partial n_c}{\partial t} = k_{ai} n_a n_a - k_{cv} n_c n_v - k_{cs} n_c n_s - v_c n_c. \tag{8.18}$$

Here k_{iv} *et al.* is the rate constant for defect denoted by subscripts (i, v, a, s and c correspond to the interstitial, vacancy, solute, sink and complex).

$$k_{iv} = 4\pi(D_i + D_v)r_{iv}; \tag{8.19}$$

$$k_{cv} = 4\pi(D_i + D_v)r_{cv}; \tag{8.20}$$

$$k_{ia} = 4\pi D_i r_{ia}. \tag{8.21}$$

D_i and D_v is the diffusion coefficient of interstitial and vacancy, respectively; r_{iv}, r_{cv}, and r_{ia} are the efficiencies of interaction of defects denoted by indexes, with each other.

$$v_c = v_{c0} \exp\left(-\frac{E_{diss}}{kT}\right). \tag{8.22}$$

v_c is the inverse lifetime of the complexes, E_{diss} is the dissociation energy of the complexes.

Since in the investigated nickel alloys the segregation is determined by output of impurity-interstitial complexes to sinks, then the effectiveness of segregation can be measured by the concentration of these complexes (n_c). The greater the complex concentration, the greater will be the segregation. By the magnitude of the relationship dn_c/dn_d (n_d is the defect concentration) can also be judged on the effectiveness of radiation-induced segregation. The experimental studies have shown that the segregation is independent of the flux density, a typical feature for the case of dominant sink absorption. At the same time, it is found that the segregation depends exponentially on fluence, i.e. the dependence observed for the solute concentration in the bulk

$$n_c \approx n_{co} \exp\left(-\frac{\Phi}{\Phi_c}\right). \tag{8.23}$$

Here Φ_c is some characteristic dose.

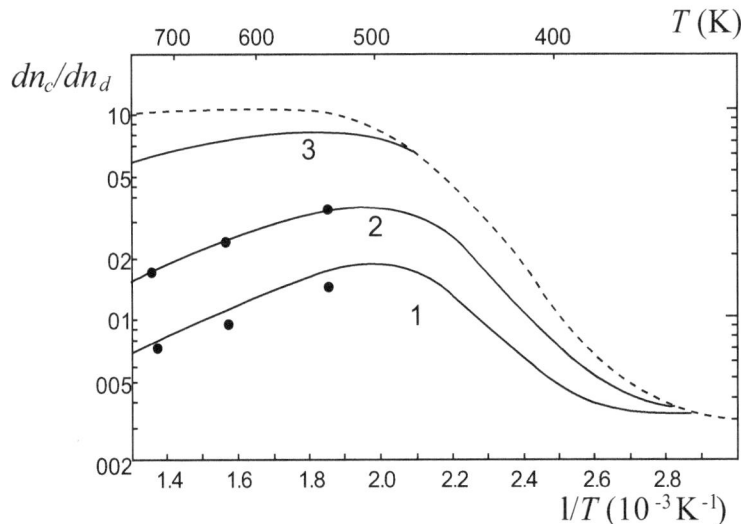

Fig. (8.6). Efficiency of the Si segregation in Ni as a function of temperature for different Si concentrations [360]. 1: 1.04 10^{-4}; 2: 2.55$10^{-4}$; 3: 20 10^{-4}.

Fig. (**8.6**) shows the experimental data and results of calculation for the Ni-Si alloy (δ_V = -0.058). At temperatures below 500 K, the efficiency of segregation decreases, since the recombination rate of point defects begins to exceed the speed of their output to the sinks. These data as well as the investigation of radiation-stimulated segregation in the Cu-Be alloy (δ_V = -0.26) provide a possibility to make a conclusion that the segregation due to the movement of complexes solute-interstitial to the sinks is more effective if the complex is more stable. In case of Ni-Ge alloy with a positive value of size misfit (δ_V = +0.15) the energy of dissociation of complex Ge-interstitial is small, the number of solute-interstitial complexes is small and the transfer of solute is realized, apparently, with participation of vacancies (due to the movement of complexes Ge-vacancy or reverse Kirkendall effect).

8.2.4. Concentrated Solid Solutions

Lam *et al.* [361] proposed the model of radiation-induced segregation for concentrated solutions. They believe that the model of radiation-induced segregation for diluted solutions cannot be used at high concentrations of solutes. The proposed model includes the following system of differential kinetic equations:

$$\frac{\partial n_v}{\partial t} = -\nabla \cdot J_v + K - R \text{,} \tag{8.24}$$

$$\frac{\partial n_i}{\partial t} = -\nabla \cdot J_i + K - R \text{,} \tag{8.25}$$

$$\frac{\partial n_a}{\partial t} = -\nabla \cdot J_a \text{.} \tag{8.26}$$

Here

$$J_v = (d_{Av} - d_{Bv})\alpha n_v \nabla n_A - D_v \nabla n_v \text{,} \tag{8.27}$$

$$J_i = -(d_{Ai} - d_{Bi})\alpha n_i \nabla n_A - D_i \nabla n_i \text{,} \tag{8.28}$$

$$J_A = -D_A \alpha \nabla n_A - n_A (d_{Ai} \nabla n_i - d_{Av} \nabla n_v) \text{,} \tag{8.29}$$

$$J_B = -D_B \alpha \nabla n_B - n_B (d_{Bi} \nabla n_i - d_{Bv} \nabla n_v) \text{,} \tag{8.30}$$

$$D_v = d_{Av} n_A + d_{Bv} n_B \text{,} \tag{8.31}$$

$$D_i = d_{Ai} n_A + d_{Bi} n_B \text{,} \tag{8.32}$$

$$D_A = d_{Av} n_v + d_{Ai} n_i \text{,} \tag{8.33}$$

$$D_B = d_{Bv} n_v + d_{Bi} n_i \text{.} \tag{8.34}$$

d_{Av}, d_{Bv}, d_{Ai} and d_{Bi} are the partial diffusion coefficients for the components of solid solution A and B by a vacancy and interstitial mechanisms of diffusion, respectively, K is the defect formation rate and R is the recombination rate. The numerical solution of this system of equations was executed for the semi-infinite geometry with parameters for the Ni-8at.%Si alloy under 400 keV He$^+$ ion bombardment. The results of calculations of solute profiles at the temperature range of 500-800 °C demonstrated good agreement with the experiment.

8.2.5. Vacancy Model of Segregation

Okamoto and Wiedersich [362] experimentally and theoretically investigated the radiation-induced segregation on the voids in stainless steel 18Cr-8Ni-1Si under electron irradiation. They suggested a simple model in which two mechanisms of segregation are proposed. The first mechanism is based on the movement of solute due to the preferential exchange of places of the substitutional solutes with vacancies. In this case the solute fluxes are directed to the opposite direction with respect to the vacancy fluxes. The second mechanism results from a strong binding between an alloying element and vacancies, which causes 'dragging' of this alloying element in the same direction as the vacancy flux. Here authors did not exclude possibility of the interstitial mechanism of segregation.

Excluding the smallest voids, the proposed model describes quite well the decrease of concentration (depletion) of chrome depending on the distance to the void under irradiation by electrons (1 MeV, ϕ = 6 10^{-4} dpa s^{-1}, Φ = 4.3

dpa). This effect becomes especially apparent for chrome. The received results are described well by the proposed expression for the solute concentration at the surface of spherically symmetrical void:

$$n_a(r) = n_a(R)\exp\left[\phi\left\{\frac{R}{r} + \frac{1}{2}\left(\frac{r}{R}\right)^2 - \frac{3}{2}\right\}\right],$$

(8.35)

$$\phi = \frac{\alpha\Phi R^2}{3D_a}.$$

(8.36)

Here $2R$ is the average distance between voids, α is the constant, Φ is the point defect generation rate taking account of the recombination, D_a is the diffusion coefficient of the solute.

8.2.6. Segregation During Irradiation by Heavy Ions

In work [363] by Johnson-Lam model has been investigated for solute segregation of substitutional solutes in dilute Ni-base binary alloys under heavy ion irradiation. The spatially dependent defect-production rates are appropriate for 75 keV and 3 MeV Ni$^+$ ion bombardment. In case of the determining interaction of solute with interstitial for example, Si in the Ni, the solute segregates from the area of maximum damages on the surface, depleting the region of peak displacements. At the temperature of approximately 600 °C the process may cause the precipitation of new phase at the surface of sinks. When the dominating complex is solute-vacancy, the surface may not only be enriched but depleted by the solute. This result depends on the temperature and binding energy of complex (E_b). If $E_b \approx 0.05$ eV, the depletion is observed on the surface and the edge of damage region, and the enrichment is observed in the peak of displacements. When $E_b > 0.1$ eV, there is the enrichment on the surface. The authors of this research work noted a good agreement of theoretical calculations with the experiment.

8.2.7. Inverse Kirkendall Effect

The inverse Kirkendall effect originates by means of vacancy fluxes to the sinks on condition of various mobilities of solute and matrix atoms. The fluxes of pointed defects and components of binary alloy **A-B** under radiation-stimulated segregation, according to the inverse Kirkendall effect mechanism, are schematically shown in Fig. (**8.7(1)**). The distributions of alloy components for various fluxes of defects are shown in Fig. (**8.7(2)**). Here n_v, n_i and $n_A{}^o$ are the concentration of vacancies, interstitial atoms and component **A** (before irradiation), respectively, J_v, J_i, $J_A{}^v$ and $J_A{}^i$ are fluxes of vacancies, interstitial atoms, **A** component according to vacancy and interstitial mechanism, respectively, x is the distance from the sink, $D_A{}^v$, $D_A{}^i$, $D_B{}^v$ and $D_B{}^i$ are coefficients of diffusion of component **A** according to vacancy and interstitial mechanism, of component **B** according to vacancy and interstitial mechanism, respectively.

The experimental data of solute distribution through the depth in 1 % alloys after their bombardment by the 3 MeV nickel ions are displayed in Fig (**8.8**) [342]. The presented data are a good example of dependence of sign and magnitude of segregation on the size interrelation of solute and solvent atoms. The undersize solute (Si) segregates the interstitial mechanism and enriches the near-surface areas of the alloy. The atoms of Al, Ni and Mo have larger size than the nickel atoms and the segregation of these solutes results in the depletion of near-surface layers. This shows the greater the difference in sizes, the higher will be the efficiency of segregation. However, inverse Kirkendall effect can only be effective in an intermediate temperature range. At low temperatures most of the Frenkel pairs are annihilated mutually because of small vacancy motion. At high temperatures (above 0.6 T_m) the concentration of thermally equilibrium vacancies throughout the volume becomes higher than the stationary concentration of radiation-induced vacancies, and the fluxes of vacancies to the sinks disappear.

Hackett *et al.* [131] using the modified inverse Kirkendall effect model, analyzed the influence of oversize solutes on the radiation-induced segregation in the Fe-Cr-Ni alloy. Their model assumed that the processes by which oversized solutes affect radiation-induced segregation are: (i) trapping, whereby freely migrating vacancies are captured by the oversized solute atom, forming a solute-vacancy complex; (ii) recombination in which freely migrating interstitials are able to recombine with the bound vacancy when the interstitial enters the recombination volume of the solute-vacancy complex; (iii) the release of vacancy by means of dissociation of complex before its recombination with interstitial atom; (iv) the migration of interstitial to grain boundaries has no effect on radiation-induced segregation. The ratio of recombination to release rate determines the effectiveness of the oversized solute in altering radiation-induced segregation.

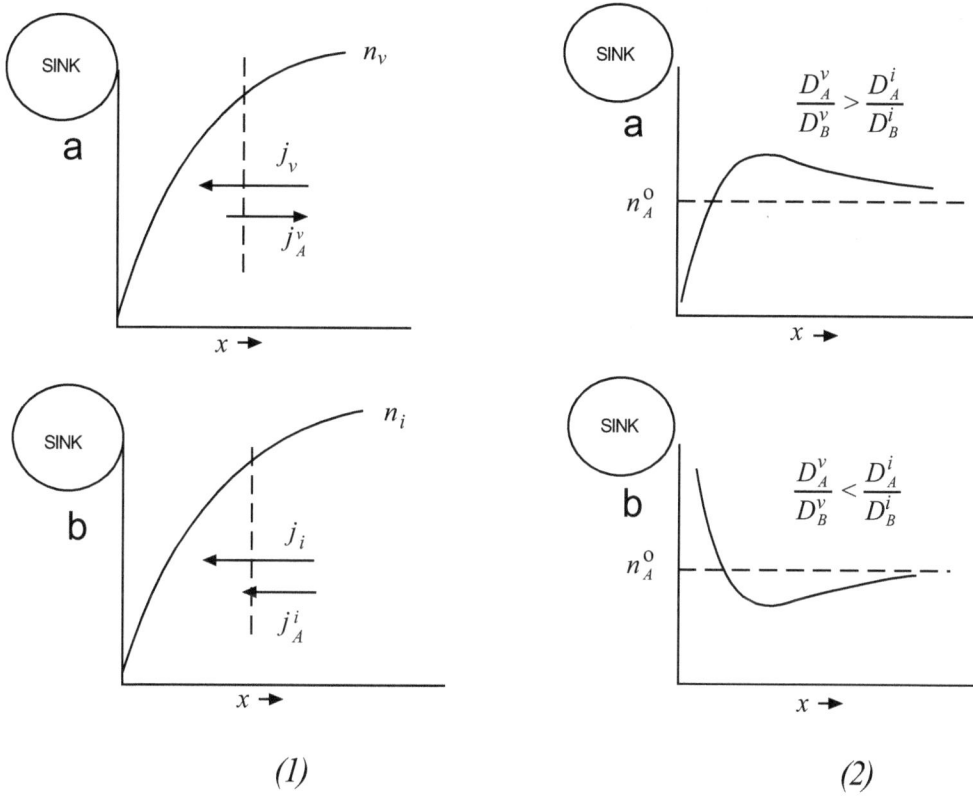

Fig. (8.7). Schematic illustration of inverse Kirkendall effect. (*1*): **a** - vacancy flux, **b** - interstitial flux; (*2*): **a** - depletion at sink, **b** - enrichment at sink [342].

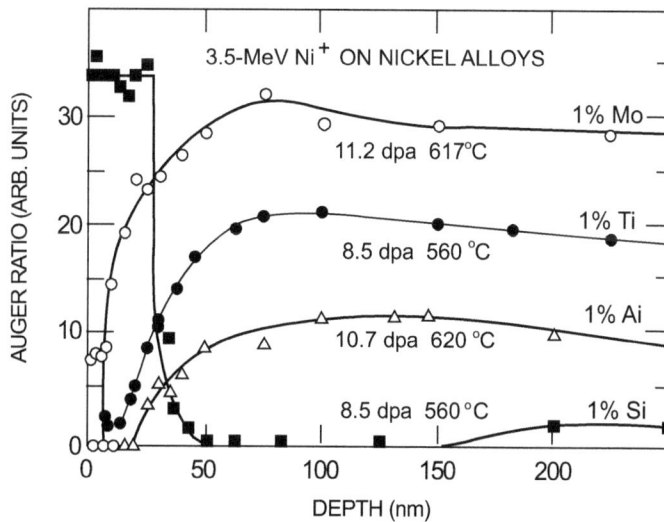

Fig. (8.8). Depth profiles of solutes (Auger ratio δ vs depth d) in a series of 1 at. % nickel alloys after ion irradiation at the temperatures and fluencies indicated [342]. 1: 1 % Si, 8.5 dpa, 560°C; 2: 1 % Mo, 11.2 dpa, 617°C; 3: 1 % Ti, 8.5 dpa, 560°C; 4: 1 % Al, 10.7 dpa, 620°C.

Unlike other models, the model of Hackett *et al.* examines not only the fluxes of defects and impurity, but the fluxes of main alloying elements Fe, Cr and Ni:

$$\frac{\partial n_{Fe}}{\partial t} = -\nabla \cdot J_{Fe},$$ (8.37)

$$\frac{\partial n_{Cr}}{\partial t} = -\nabla \cdot J_{Cr},$$ (8.38)

$$\frac{\partial n_{Ni}}{\partial t} = -\nabla \cdot J_{Ni},$$ (8.39)

$$\frac{\partial n_v}{\partial t} = -\nabla \cdot J_v + K - k_{iv}(D_i + D_v)n_i n_v - \rho_v D_v (n_v - n_v^{th}) - k_- n_s n_v + k_c^- n_c,$$ (8.40)

$$\frac{\partial n_i}{\partial t} = -\nabla \cdot J_i + K - k_{iv}(D_i + D_v)n_i n_v - \rho_i D_i n_i - k_{ci} n_c n_i.$$ (8.41)

Here n_v, n_i, n_a and n_c are the concentrations of free vacancies, interstitials, impurity atoms and impurity-vacancy complexes, respectively, n_v^{th} is the thermally equilibrium concentration of vacancies, $\nabla \cdot J_x$ is the divergence of flux of alloying elements and defects (x represents either the alloying elements or vacancies or interstitials), K is the generation rate of pointed defects , k_{iv}, k_c, k_{ci} and k_c^- are the constants of recombination rate of free vacancies with interstitial atoms, of vacancy capture by an impurity atom and release of vacancies, respectively, D_v and D_i are the diffusion coefficient of vacation and interstitial atoms, respectively, ρ_v and ρ_I are the density of sinks for vacancies and interstitial atoms, respectively.

The considered inverse Kirkendall effect model also contains rate equations for the concentration of solute-vacancy complexes and unbound solute atoms:

$$\frac{\partial n_c}{\partial t} = k_c n_a n_v - k_c^- n_c - k_{ci} n_a n_i,$$ (8.42)

$$\frac{\partial n_a}{\partial t} = -k_c n_a n_v + k_c^- n_c + k_{ci} n_a n_i.$$ (8.43)

The set of equations was solved numerically to obtain concentration profiles of solutes as a function of time and distance from the grain boundary. The calculations were performed for the initial solute concentration of 10^{-3} and $K = 10^{-6}$ dpa s^{-1}. The main conclusion from the performed calculations is the fact that the addition of oversize solute

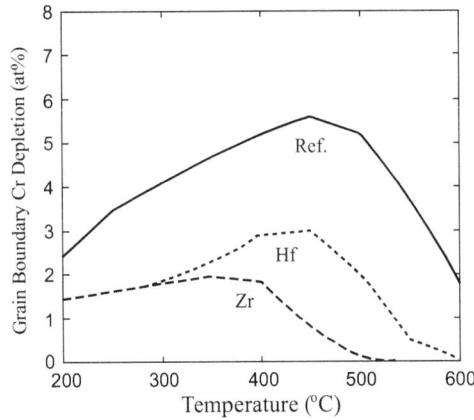

Fig. (8.9). Grain boundary Cr (Δn_{Cr}) depletion for the inverse Kirkendall effect model as a function of temperature for Fe–15Cr–13Ni and with 0.1 at. % Zr or Hf additions at $\phi = 10^{-6}$ dpa s^{-1} and $\Phi = 1$ dpa.

into the austenitic stainless steel Fe-15Cr-13Ni may result in a significant reduction in radiation-induced segregation of chrome. Among four analyzed solutes Zr, Hf, Ti and Pt, the highest effect was observed in the first two. This is demonstrated in Fig. (**8.9**), where the values of Cr depletion at the grain boundary without and with Zr and Hf depend on the irradiation temperature. The effectiveness of the radiation stimulated segregation depends apparently on the difference in the solute-vacancy binding energy. The higher the binding energy, the more will be the effect.

8.2.8. Model of Elastic Interaction of Point Defects with Sinks

Sorokin and Ryazanov [364] analyzed the effect of elastic stress fields caused by sinks (grain boundaries) on the radiation-induced segregation in the binary alloys. They considered the grain boundary as a row of dislocations, uniformly distanced from each other at a distance (h). The energy of elastic interaction of vacancy or interstitial with grain boundary will be proportional to the spherical dilatation of point defect (δ_d), to the Burger's vector (b) and the shear module (μ):

$$E_d = C\mu b\delta_d . \tag{8.44}$$

Here C is the coefficient which has a complicated trigonometric dependence from h. The calculations demonstrated that the elastic interactions cause the decrease of concentration of pointed defects near the sinks, as well as the decrease of critical temperature of segregation T_c, if $E_v < 0$ ($E_i > 0$) or to growth T_c, if $E_v > 0$ ($E_i < 0$).

CHAPTER 9

Radiation-Induced Disordering and Amorphization

Abstract: The radiation-induced disordering and amorphization are considered. These phenomena are not connected with the fluxes of impurity atoms and radiation defects. The criteria for radiation amorphization which is connected with the degree of mixing of the alloy components, are discussed.

Keywords: Radiation, metal, alloys, solid solution, disordering, amorphization, radiation amorphization models, criteria of radiation amorphization.

This chapter analyzes the radiation-stimulated phenomena that are not connected with the fluxes of radiation defects and solute atoms. Among such phenomena, which are realized only by means of restructuring of crystal lattice, are radiation-induced ordering, disordering and phase transformations of one crystal structure to the other or escape from the crystal state to the amorphous one and vice versa.

9.1. RADIATION-INDUCED ORDERING

For the binary solution consisting of atoms **A** and **B**, the ordering corresponds to an increase in the probability that the **A**-atom occupies an α-site, and that the **B**-atom occupies β-site. The characteristic feature of radiation ordering is the fact that in some cases a very small dose is required for its realization. For example, for the ordering of rapidly solidified Sn–6.7Sb–5.3Zn alloy under room temperature, the γ-radiation takes place with the dose of 50 Gy (^{60}Co, P_d = 5 Gy s^{-1}) [365]. The small dose under irradiation by electrons (1-3 MeV, $\Phi \approx 10^{-5}$ dpa) also results in the ordering of Cu$_3$Au, Ni$_3$Mn, CuZn alloys [366-369]. The amazing element in the last case is the ordering of already ordered alloy.

During the ordering, the electrical resistivity of alloys decreases. This provides a possibility to use the method of measurement of electrical resistivity for controlling the process of ordering or disordering. Using this method, Blewitt and Coltman [370] determined that the irradiation by fast neutrons (2 10^{-4} dpa) of the disordered Cu$_3$Au at 113 K doesn't change the resistivity. At the same time, with the same fluence of neutrons at 423 K, the resistivity falls to 22 %. An elevated temperature is an important condition for the occurrence of radiation-induced ordering [59]. The same authors showed that the ordering of disordered Cu$_3$Al increases during fast-neutron irradiation at 473 K, at a rate which is about four orders of magnitude greater than that expected in the absence of irradiation [371].

As a result of exploring radiation-enhanced ordering in the short-range in the α-Ag-Zn alloy, Halbwachs and Hillairet [372] came to the unexpected conclusion. According to their conclusion, the mobility of self-interstitials (silver) is lower than the mobility of vacancies in the explored alloy. When processing the experimental data they used usual balance equations of point defects.

Based on evidences gathered we may make a conclusion that the formation and the thermally activated migration of radiation vacancies is a dominating mechanism for ordering the solid solution under irradiation.

9.2. RADIATION-INDUCED DISORDERING

Siegel [373] one of the first who showed that irradiation with neutrons of ordered alloy leads to disordering. According to his data, after irradiation by fast neutrons (3.3 10^{23} m^{-2}) at 310 K, the electrical resistivity of ordered Cu$_3$Al alloy increases to the value which is close to the same one for the disordered alloy. X-ray structural analysis showed that the initially ordered alloy was partially disordered.

Howe and Rainville [374] experimentally studied the radiation disordering in the Zr$_3$Al under irradiation by 0.5-2.0 MeV Ar$^+$ ions. It was found that at low ion fluencies (up to 10^{16} m^{-2}) individual damage regions were observed. At ion fluencies 10^{16}-10^{18} m^{-2} overlap of the damage regions occurred thus resulting in a complex damage configuration. With higher fluencies (up to 10^{20} m^{-2}) the structure of Zr$_3$Al transfers to the amorphous state.

Recrystallization from the amorphous state occurred during post-irradiation annealing at temperatures above 723 K for amorphous material present on an underlying crystalline matrix and the full amorphisation occurs at temperatures 973 K.

From the data presented by Schulson [375] it is clear that at a temperature below 300 K the damage required to essentially disorder (from $S = 1$ to $S = 0.1$, S is degree of order) any of the alloys of A_3B type under fast-neutron irradiation is equivalent to around 0.1 to 0.3 dpa. For the same disorder under irradiation of 1 MeV electrons the required dose is an order of magnitude higher than 1-3 dpa, for Zr_3Al and Ni_3Al.

According to data of Carpenter and Schulson [376] the dose of 1 dpa causes full randomization in the Zr_3Al under irradiation by the 1 MeV electrons at the temperature of 130-375 K. At these temperatures, the rate of electron induced disordering doesn't depend on dose rate, at least in the range $5 \cdot 10^{-4}$-$5 \cdot 10^{-3}$ dpa s^{-1}. At the temperatures of 575-775 K, the degree of long-range order is not stable and has an intermediate value. The number of atomic displacements to cause complete disordering at the low temperatures appears to be the same for both energetic ion and energetic electron damage. Electron irradiation at low temperatures to doses in excess of about one dpa produces damaged regions in which the crystalline perfection is greatly reduced. It should be noted that Zr_3Al is resistant to void formation.

Schulson [375] considers five mechanisms to account for the phenomenon of radiation disordering: thermal spikes by Seitz; replacement collisions by Kinchin and Pease; plastic spikes by Seitz and Koehler; collapse of cascades to vacancy/interstitial recombination by Jenkins and Wilkens; and random vacancy/interstitial recombination by Carpenter and Schulson. The spike lifetime is either of the order of magnitude or longer than the period of cascade production, suggesting that thermally-activated diffusion may account for the atomic rearrangement. The mechanism of replacement collision sequences is quite probable. Calculations have shown [377] that one recoil atom with an energy of 450 eV causes about 132 replacements in Ni_3Mn. Disordering by the plastic spike mechanism is unlikely since in the disordered areas the dislocations are not found. Random recombination mechanism, although possible, could not account for the large number of the replacements.

Of all the above mechanisms the most likely are the thermal spikes and the replacement collision sequences. The predominant mechanism of disordering is dependent upon the nature of the irradiation. Under irradiation by ions, probably the mechanism of thermal spikes and the replacement collisions with long chains, will be realized. In case of electron irradiation, disordering by means of random recombination of interstitials with vacancies and replacement collisions with short chains must dominate.

9.3. BANERJEE-URBAN'S MODEL OF RADIATION-INDUCED DISORDERING

Banerjee and Urban [378] proposed a model of radiation-induced disordering of solid solutions, based on the Bragg-Williams approximation. It is assumed that radiation-induced disordering is opposed by radiation-enhanced reordering. The order-disorder transformation is realized by means of jumps of thermal and radiation vacancies. If the vacancy changes sublattice, the jumping atom may perform either an "ordering" or a "disordering" jump dependent on whether it eventually occupies the "right" or "wrong" sublattice. The diffusion of interstitials practically doesn't affect on the degree of order:

$$S = \frac{p_A^\alpha - n_A}{1 - n_A} = \frac{p_B^\beta - n_B}{1 - n_B} . \tag{9.1}$$

It is supposed that the activation energy of jumps (U) is a linear function of the degree of order:

$$U = 2(z_{\beta\alpha} - z_{\alpha\alpha})ES , \tag{9.2}$$

here

$$E = E_{AB} - \frac{E_{AA} + E_{BB}}{2} . \tag{9.3}$$

Indexes of α and β denote the sites of sublattices of **A** and **B** atoms, $z_{\beta\alpha}$ and $z_{\alpha\alpha}$ are the coordination numbers, $p_A{}^{\alpha}$ and $p_B{}^{\beta}$ are the probabilities of an **A**-atom residing on an α-site and a **B**-atom residing on a β-site, respectively, and n_A and n_B are the atomic fractions of the respective species. For the calculation the following system of equation can be used:

$$\frac{dn_v^{\alpha}}{dt} = K(1 - mn_v^{\alpha}) - k_r n_i n_v^{\alpha} + k_{\beta\alpha}^{\alpha}(n_v^{\beta} - n_{th}^{\beta}) - k_{\alpha\beta}^{\alpha}(n_v^{\alpha} - n_{th}^{\alpha})n_s, \tag{9.4}$$

$$\frac{dn_v^{\beta}}{dt} = K(1 - mn_v^{\beta}) - k_r n_i n_v^{\beta} + k_{\alpha\beta}^{\beta}(n_v^{\beta} - n_{th}^{\beta}) - k_{\beta\alpha}^{\beta}(n_v^{\alpha} - n_{th}^{\alpha})n_s, \tag{9.5}$$

$$\frac{dn_i}{dt} = K(1 - mn_v) - n_i(k_r^{\alpha} + k_r^{\beta}) - k_{is}n_i n_s, \tag{9.6}$$

where

$$k_r = \left(\frac{4\pi r_0}{\Omega}\right)D_i,$$

$k_r^{\alpha} = n_A k_r$, $k_r^{\beta} = n_B k_r$. The values for other parameters of the system of equations are given in reference [378]. The disordering rate during irradiation is a sum of three components:

$$\frac{dS}{dt} = \left(\frac{dS}{dt}\right)_c + \left(\frac{dS}{dt}\right)_r + \left(\frac{dS}{dt}\right)_{th}. \tag{9.7}$$

Here disordering rate due to collision sequences is

$$\left(\frac{dS}{dt}\right)_c = -\frac{\chi m_e KS}{n_A n_B}. \tag{9.8}$$

The factor χ takes into account that depending on the alloy structure, there may be close-packed lattice directions of a certain type for which not all atom rows exhibit an alternating occupation with α and β sites. If l is the length of replacement collision sequence, p is the periodicity of α and β sites, then

$$m_e = \frac{l}{p}\left[1 - mn_v\frac{l-1}{2l}\right]. \tag{9.9}$$

Disordering rate by random radiation defect annihilation is

$$\left(\frac{dS}{dt}\right)_r = KS. \tag{9.10}$$

Disordering rate by thermal vacancy motion is

$$\left(\frac{dS}{dt}\right)_{th} = k_+ n_A^{\beta} n_B^{\alpha} - k_- n_A^{\alpha} n_B^{\beta}. \tag{9.11}$$

k_+ and k_- are the ordering and disordering rate constants, respectively. Using the model described above, the calculations for the hypothetic **AB**$_4$ alloy were executed. In order to simplify comparisons with experimental results, received for the alloy Ni$_4$Mo, the values of parameters were chosen in accordance with literature data for pure nickel. Fig. (**9.1**) presents the calculated dependence of S value from the irradiation temperature for $\phi = 5 \cdot 10^{-3}$ dpa s^{-1} and $E_m = 1.38$ eV, as well as the experimental data for the Ni$_4$Mo [379]. It is obvious these results and the calculation are in good agreement with each other.

9.4. RADIATION-INDUCED AMORPHIZATION

Holz *at al.* [380] found that the bombardment by heavy ions (275 kcV Ar$^+$, 250 keV Ne$^+$) at the temperatures lower than 10 K can result in a complete amorphization of α-Ga. It is also determined that with the increase of absorbed

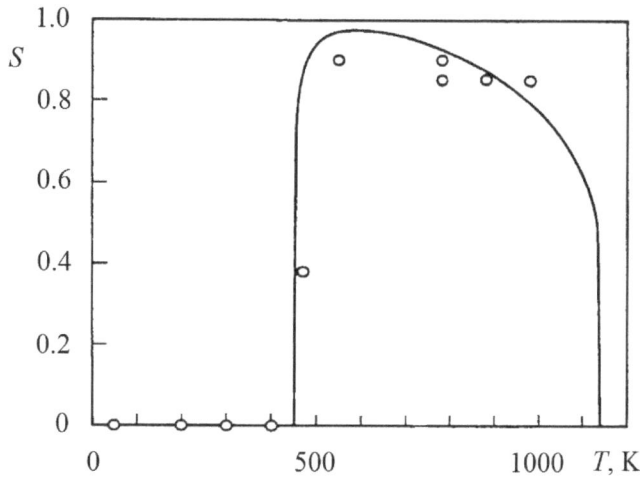

Fig. (9.1). The calculated temperature dependence of the steady-state degree of long-range order, *S*, for $\phi = 5 \cdot 10^{-3}$ dpa s^{-1} and E_m = 1.38 eV in the Dl$_a$ alloy [378]. The data points are from electron irradiation experiments on Ni$_4$Mo [379].

energy the relation T_{cs}/T_{ca} grows from 0.1 to 1 (here T_{cs} is the temperature of transfer in the condition of superconductivity, T_{ca} is the irradiation temperature with achieved amorphization). It is important to note that irradiation by helium ions with an equivalent absorbed dose, does not lead to amorphization. This points to the fact that the radiation amorphization needs a high value of linear transmission of energy. It is necessary to note that under identical irradiation conditions, amorphization in β-Ga will not occur.

During the investigation of radiation disordering in the nickel-manganese alloy Aronin [381] found that the degree of long-range order is situated in the exponential dependence on the absorbed dose (ϕt)

$$S = S_0 \exp(-A\phi t),\qquad\qquad\qquad (9.12)$$

where S_0 is the initial long-range order parameter, *A* is a parameter which is a measure of the effectiveness with which the incident particle disorders the material.

Sweedler and Cox [74] also observed an exponential decrease in *S* with the increase of fluence for the Nb$_3$Al alloy, irradiated at 140 °C with fast neutron to $\phi = 5 \cdot 10^{23}$ m^{-2}.

Jaouen *et al.* [382] presented the results of experimental investigation of the impact of energy of incident ions on the transformation from the crystalline to amorphous state in the Ni-Al alloy. The various ions (^{209}Bi^{3+}, ^{129}Xe^{2+}, ^{40}Ar^{2+}, ^{40}Ar$^+$, ^{4}He$^+$), energies (540 keV, 360 keV, 180 keV, 120 keV), fluences (1.4 10^{16}-0.2 10^{16} m^{-2}) and the number of displacements per ion (4415-5) were used. On the basis of performed researches the authors made the following conclusions: (i) transformation of crystal state in the amorphous is possible only inside the cascade created by a heavy ions; (ii) light ions cannot cause amorphization at the temperature of 77 K, and they create only the partial disordering; (iii) the local density of displacements or the energy density play the important role in the structure transformations; (iv) accumulation of point defects, apparently, does not impact on the transformation; the effectiveness of amorphization and the defect formation stay inversely from the energy density.

Radiation-induced transformation of the crystalline-amorphous state in the iron with boron has its own features. Amorphization of thin layers of iron (50-60 nm) with the implantation of boron (up to 27 keV, 3 10^{21} m^{-2}) at room temperature occurs only when a result is achieved by implantation of the eutectic composition [383].

For some other compounds the radiation-induced amorphization is possible not only during irradiation with heavy ions. For example, amorphization of the memory Ti-Ni alloy is observed as a result of irradiation by fast neutrons with fluences larger than 1.2 10^{23} m^{-2} [384].

Parsons [385] proposed the following mechanism of radiation amorphization. This mechanism is characterized by discrete amorphous zones which are produced directly within certain collision cascades by the rapid cooling from the very high spike temperatures at which localized melting occurs. This mechanism implies that the amorphous zones are formed at the beginning of irradiation.

9.5. CRITERIA OF RADIATION AMORPHIZATION

Naguib and Roger Kelly [386] being based on the data for non-metal material, proposed that a temperature criterion states that if the crystallization temperature of the amorphous phase exceeds 0.3 of the melting point then the material should become amorphous under irradiation.

According to the other criterion the amorphization must take place if the degree of binding iconicity (χ) is less than the value of 0.47. Here $\chi = \exp[-0.25(x_A - x_B)^2]$, x_A and x_B are electronegativities. For metal alloys $\chi < 0.1$ and this criterion apparently does not work. On the other hand, the existence of amorphous transformation in the Ni-Al$_3$ alloy and the absence of radiation amorphization in the Ni$_3$Al alloy, possibly confirms that the impact of degree of binding iconicity exists, since the bonding in Ni-Al$_3$ is more covalent than the Ni$_3$Al.

Brimhall *et al.* [387] presented the results of experimental research of various intermetallic compositions in order to get the general criteria for amorphous transformation in metallic alloys during irradiation. The combinations of transition metals and alloys, including aluminum served as samples. All the alloys were bombarded with 2.5 MeV ions Ni$^+$ at the room temperature. The analysis of received results demonstrated that the ability for radiation amorphization correlated well with the degree of solubility, except for certain cases. The compounds in which the solubility is less than about 1 % (Ni-Ti, Ni-Ti$_2$, Ni-Al$_3$, Fe-Ti, Fe2-Ni, Mo-Ni), have shown greater tendency for amorphous transformation at dose $D \leq 1$ dpa. The compounds with a wide range of solubility (Ni$_3$-Ti, Ni-Al, Ni$_3$-Al, Fe-Al, Fe$_3$-Al) did not undergo the amorphous transformation even if the dose is higher than 10 dpa.

The convincing explanation of radiation-induced amorphous transformation is given by Swanson et al. [388]. They proposed that the transfer from crystalline to amorphous state takes place when the full energy of radiation defects (E_d) in the solid exceeds the energy of amorphous transformation without irradiation (E_{am}). To E_d there corresponds some critical concentration of radiation defects (n_{am}). For example, for germanium $n_{am} \approx 0.02$ [388]. According to Brimhall *et al.* [387] for equivalent radiation doses, compounds with limited solubility will undergo a greater increase in free energy than those with wide compositional range. It is conditioned by the fact that such compositions cannot exist in the thermodynamical equilibrium out of the solubility range. At temperatures above room temperature, the steady-state concentration of radiation defects decreases with increasing temperature. For this reason the amorphous transformation in compounds with high value n_{am} is possible only at low irradiation temperatures.

Analysis of experimental data, executed by Brimhall *et al.* [381] provides a possibility to think that alloys with small solubility limit (< 1 %) are susceptible to radiation amorphisation, and compounds with a wide range of mixing do not undergo conversion of crystal-amorphous state under irradiation.

9.6. RADIATION-INDUCED PHASE TRANSFORMATION

In the work [389] it is demonstrated that a compositionally modified by 12 % Cr, steel, containing a 4.6 % Ni addition and with initially fully martensitic structure, undergoes partial transformation to austenite, following high dose neutron irradiation. Transmission electron microscopy investigation shows that the fully martensitic weld metal transformed to a duplex austenite/ferrite structure due to fast neutron irradiation (1.7 10^{-6} dpa s^{-1}, 28 dpa) at 465 °C. The austenitic was heavily voided (about 15 vol. %) and the ferrite was relatively void-free. In order to explain such a significant decrease of temperature of phase transfer (on 450 °C) from bcc to fcc structure, the following mechanism is proposed. Existence of the dislocation loop nucleus decreases the energetic barrier of interphase transformation. Calculations have shown that the observed temperature decrease of the interphase alpha-gamma transition takes place with increasing density of dislocation loops up to 3 10^{21} m^{-3} with the size of 35 nm. The energy reserved due to the formation of dislocation loops amounts to 1030 J mol^{-1}. The change in free energy of the system due to the formation of dislocation loops may be estimated according to the following expression:

$$G_i^{irr} = 2\pi R \frac{\mu b^2}{4\pi(1-v)} \left(\ln \frac{4R}{r_0} - 1 \right) \rho_l, \tag{9.13}$$

Where R is the radius of the circular dislocation loop, μ is the shear modulus of the bulk material, b is the magnitude of the dislocation Burgers vector, v is the Poisson's ratio, r_0 is a dislocation core cut-off radius, and ρ_l is the dislocation loop density.

REFERENCES

[1]   Hayashi, N.; Toriyama, T.; Sakamoto, I.; Hisatake, K. The effect of 40 keV helium ion irradiation on Fe-B amorphous alloys. *J. Phys. Condens. Matter.*, **1989**, *1*, 3849-3858.

[2]   Lam, N.Q.; Janghorban, K.; Ardell, A.J. On the modeling of irradiation-induced homogeneous precipitation in proton-bombarded Ni-Si silis solutions. *J. Nucl. Mater.*, **1981**, *101*, 314-325.

[3]   Basu, A.M.; Choudhury, A.; Chaudhuri, S.; Mukhopadhyay, P.; Baliga, B.B. The influence of impurities on the range of 28 MeV α-particles in NaCl single crystals. *Phys. Stat. Sol. (b)*, **1985**, *128*, 483-488.

[4]   Ehrhard, P.; Schlagheck, U. Investigation of Frenkel defects in electron irradiated copper by Huang scattering of x rays: I. Results for single interstitials. *J. Phys. F*, **1974**, *4*, 1575-1588.

[5]   Bender, O.; Ehrhart, P. Self-interstitial atoms, vacancies and their agglomerates in electron-irradiated nickel investigated by diffuse scattering of x-rays. *J. Phys. F*, **1983**, *13*, 911-928.

[6]   Sangster, M.J.L.; Rowell, D.K. Direct calculation of force constant changes around point defects in crystals: substitutional impurities, vacancies and F centers. *J. Phys. C*, **1982**, *15*, 5153-5159.

[7]   Siegel, R.W. Vacancy concentration in metals. *J. Nucl. Mater.*, **1978**, *69-70*, 117-146.

[8]   Varotsos, C.; Lazaridou, M.; Alexopoulos, K.; Varotsos, P. Point defect entropies and enthalpies in KCl. *Phys. Stat. Sol. (b)*, **1985**, *130*, K105-107.

[9]   Arends, J. Defect formation energies and migration energy barriers in cesium halides. *Phys. Stat. Sol (b)*, **1967**, *24*, K129-131.

[10]   Simmons, R.O.; Balluffi, R.W. Measurement of equilibrium vacancy concentrations in aluminum. *Phys. Rev.*, **1960**, *117*, 52-61.

[11]   Simmons, R.O.; Balluffi, R.W. Measurement of equilibrium concentrations of lattice vacancies in gold. *Phys. Rev.*, **1962**, *125*, 862-872.

[12]   Janot, Chr.; George, B. Equilibrium concentration of vacancies in hexagonal metals. *Phys. Rev. B*, **1975**, *12*, 2112-2219.

[13]   Simmons, R.O.; Balluffi, R.W. Measurement of equilibrium concentrations of vacancies in copper. *Phys. Rev.*, **1963**, *129*, 1533-1544.

[14]   Rasch, K.-D.; Siegel, R.W.; Schultz, H. Quenching and recovery experiments of tungsten. *J. Nucl. Mater.*, **1978**, *69-70*, 622-624.

[15]   Hautojarvi, P.; Pollanen, L.; Vehanen, A.; Yli-Kauppila, J. Vacancies and carbon impurities in α-iron: neutron irradiation. *J. Nucl. Mater.*, **1983**, *114*, 250-259.

[16]   Van der Kolk, G.J.; Post, K.; Van Veen, A.; Pleiter, F.; De Hosson, J.Th.M. Interaction of vacancies with implanted metal atoms in tungsten observed by means of thermal helium desorption spectrometry and perturbed angular correlation measurement. *Rad. Eff.*, **1985**, *84*, 131-158.

[17]   Trushin, Yu.V. *Physical Basis of Radiation Material*; SPU: St. Petersburg, **1996**.

[18]   Schüle, W. Vacancy enhancement of diffusion after quenching and during irradiation in silver-zinc alloys. *J. Phys. F*, **1980**, *10*, 2345-2357.

[19]   Pershits, Ya.N.; Veisman, V.L. Determination of point defect energy parameters in NaBr crystal using one and two-valent impurities. *Fiz. Tverd. Tela*, **1983**, *25*, 1379-1385.

[20]   Schilling, W. Self-interstitial atoms in metals. *J. Nucl. Mater.*, **1978**, *69-70*, 465-489.

[21]   Bender, O.; Ehrhart, P. Self-interstitial atoms, vacancies and their agglomerates in electron-irradiated nickel investigated by diffuse scattering of x-rays. *J. Phys. F*, **1983**, *13*, 911-928.

[22]   Dimitrov, C.; Moreau, F.; Dimitrov, O. Influence of magnesium additions on the low temperature recovery stages in neutron irradiated aluminium. *J. Phys. F*, **1975**, *5*, 385-399.

[23]   Rowell, D.K.; Sangster, M.J.L. Calculations of intrinsic defect energies in the alkali halides. *J. Phys. C: Solid State Phys.*, **1981**, *14*, 2909-2921.

[24]   Little, E.A. J. Void-swelling in irons and ferritic steels: I. Mechanisms of swelling suppression. *J. Nucl. Mater.*, **1979**, *87*, 11-24.

[25]   Sild, O. The model of rotation of the NO_2^- impurity molecule in KCl crystal. *Phys. Stat. Sol. (b)*, **1990**, *161*, 105-111.

[26]   Maes, F.; Callens, F.; Matthys, P.; Boesman, E. EPR of S_2^- in NaBr. *Phys. Stat. Sol. (b)*, **1990**, *161*, K1-3.

[27]   Banasevich, S.N.; Lurje, B.G.; Murin, A.N. Determination of effective charge of Ca-ions in the compound crystals of $NaCl^+CaCl_2$. *Fiz. Tverd. Tela*, **1960**, *2*, 80-87.

[28]   Tsuboi, T.; Polosan, S.; Topa, V. The Pb^{2+}(Li) and Pb^{2+}(Na) centres in KCl crystals. *Phys. Stat. Sol. (b)*, **2000**, *217*, 975-980.

[29]   Nagirnyi, V.; Stolovich, A.; Zazubovich, S.; Janson, N. Time-resolved polarization spectroscopy and off-centre position of excited Ga^+ ions in NaCl:Ga crystal. *Phys. Stat. Sol. (b)*, **1992**, *173*, 743-753.

[30]   Ohgaku, T.; Takeuchi, N. Interaction between a dislocation and monovalent impurities in KCl single crystals. *Phys. Stat. Sol. (a)*, **1992**, *134*, 397-404.

[31]   Cottrell, A.; Bilby, B. Dislocation theory of yielding and strain ageing of iron. *Proc. Phys. Soc. A*, **1949**, *62*, 49-61.

[32]   Andreev, G.A.; Buritskova, L.G. Linear growth of two-phase regions during unmixing of NaCl-KCl crystals. *Cryst. Res. Technol.*, **1984**, *19*, 155-163.

[33]   Denisov, E.T. *Kinetics of Homogenous Chemical Reaction*; "Vysshaya Shkola": Moscow, **1978**.

[34]   Suzuki, K. X-ray studies on precipitation of metastable centers in mixed crystals $NaCl-CdCl_2$. *J. Phys. Soc, Jap.*, **1961**, *16*, 67-78.

[35]   Sors, A.I.; Lilley E. Anion displacements and the lattice energy of the 6 $NaCl:MCl_2$ family of structures. *Phys. Stat. Sol. (a)*, **1975**, *27*, 469-475.

[36]   Cywinski, R. Fluorescence in doubly doped $KCl:Eu^{2+}$, Sm^{2+} crystals. *Phys. Stat. Sol. (b)*, **1994**, *186*, 575-580.

[37]   Cywinski, R.; Mugenski, E.; Nowy-Wiechula, W.; Wiechula, J. Suzuki phase precipitates in doubly doped $KCl:Pb^{2+}$, Mn^{2+} crystals. *Phys. Stat. Sol. (b)*, **1989**, *151*, 47-51.

[38]   Babin, V.; Fabeni, P.; Mihokova, E.; Nikl, M.; Pazzi, G.P.; Zazubovich, N.; Zazubovich, S. Luminescence of Cs_4PbBr_6 aggregates in as-grown and in annealed CsBr:Pb single crystals. *Phys. Stat. Sol. (b)*, **2000**, *219*, 205-214.

[39]   Cywinski, R.; Mugenski, E; Nowy-Wiechula, W.; Wiechula, J. Agregation of Mn^{2+}-cation vacancy dipoles in KCl matrix. *Phys. Stat. Sol (b)*, **1985**, *129*, 601-605.

[40]   Kirk, M.A.; Blewitt, T.H. Atomic rearrangements in ordered fcc alloys during neutron irradiation. *Met. Trans.*, **1978**, *9A*, 1729-1737.

[41]   Shalaev, A.M. *Radiation-Induced Processes in Metals*. "Energoatomizdat": Moscow, **1988**.

[42]   Daou, J.N.; Vajda, P.; Lucasson, A.; Lucasson, P. Electron radiation damage in the rare-earth metals yttrium, samarium and ytterbium at low temperatures. *J. Phys. F*, **1980**, *10*, 583-597.

[43]   Terentyev, D.A.; Malerba, L.; Chakarova, R.; Nordlund, K.; Olsson, P.; Rieth, M.; Wallenius, J. Displacement cascades in Fe–Cr: A molecular dynamics study. *J. Nucl. Mater.*, **2006**, *349*, 119-132.

[44]   Kiritani, M. Point defect processes in the defect structure development from cascade damage. *Mater. Sci. Forum*, **1987**, *15-18*, 1023-1046.

[45]   Katsnelson, A.A.; Goncharenko, Yu.D.; Grabova, R.B.; Kosenkov, V.M. X-ray studies of neutron irradiated metals with bcc and fcc lattices. *Fiz. Metal. i Metalloved.*, **1984**, *57*, 138-141.

[46]   Kozlov, A.V. Dependence of the concentration of point defects in the ChS-68 austenitic steel on the rate of their generation and temperature upon neutron irradiation. *Phys. Metal. Metallog.*, **2009**, *107*, 534-541.

[47]   Piller, R.C.; Marwick, A.D. Radiation-enhanced diffusion and segregation in Ni and Ni(Si) alloy: the effects of point-defect trapping. *J. Nucl. Mater.*, **1979**, *83*, 42-47.

[48]   Brailsford, A.D.; Bullough, R. The rate theory of swelling due to void growth in irradiated metals. *J. Nucl. Mater.*, **1972**, *44*, 121-135.

[49]   Ward, A.E.; Fisher, S.B. Dislocation loop growth in pure iron under electron irradiation. .*J. Nucl. Mater.*, **1989**, *166*, 227-234.

[50]   Sekimura, N.; Shirao, Y.; Yamaguchi, H.; Yonamine, S.; Arai, Y. Defect cluster formation in vanadium irradiated with heavy ions. *J. Nucl. Mater.*, **1999**, *271-272*, 63-67.

[51]   Hishinuma, A.; Katano, Y.; Fukaya, K.; Shiraishi, K. Re-irradiation effect on nucleation of voids in stanless steel. *J. Nucl. Mater.*, **1975**, *55*, 227-228.

[52]   Lebedev, V.M.; Lebedev, V.T.; Ivanova, I.N.; Orlov, S.P.; Orlova, D.N. Structure of aluminum alloys irradiated with reactor neutrons. *Phys. Sol. State*, **2010**, *52*, 996-999.

[53]   Brimhall, J.L.; Kulcinski, G.L. Void formation in ion bombarded niobium. *Rad. Eff.*, **1973**, *20*, 25-31.

[54]   Barashev, A.V.; Golubov, S.I. On the onset of void ordering in metals under neutron or heavy-ion irradiation. *Phil. Mag.*, **2010**, *90*, 1787-1797.

[55]   Chen, L.J.; Ardell, A.J. Void ordering in nitrogen-ion nickel-aluminum solid solution. *J. Nucl. Mater.*, **1978**, *75*, 177-185.

[56]   Yang, W.J.; Dodd, R.A.; Kulcinski, G.L. Electron irradiation damage in high purity aluminum. *J. Nucl. Mater.*, **1977**, *64*, 157-166.

[57]   Bae, D.S.; Nahm, S.H.; Lee, H.M.; Kinoshita, H.; Shibayama, T.; Takahashi, H. Effect of electron-beam irradiation temperature on irradiation damage of high Mn–Cr steel. *J. Nucl. Mater.*, **2004**, *329-333*, 1038-1042.

[58]   Rauch, R.; Peils, J.; Schmalzbauer, A.; Willner, G. Loop formation in Cu and Al after low-temperature fast-neutron irradiation. *J. Phys.: Condens. Matter*, **1990**, *2*, 9009-9017.

[59]   Gilbert, R.W.; Holt, R.A. Dislocation structure in neutron irradiated zircaloy. *J. Nucl. Mater.*, **1981**, *102*, 1-6.

[60]   Okita, T.; Sato, T.; Sekimura, N.; Garner, F.A.; Greenwood, L. R. The primary origin of dose rate effects on microctructural evolution of austenitic alloys during neutron irradiation. *J. Nucl. Mater.*, **2002**, *307-311*, 322-326.

[61]   Okita, T.; Sato, T.; Sekimura, N.; Iwai, T.; Garner, F.A. The synergistic influence of temperature and displacement rate on microstructural evolution of ion-irradiated Fe–15Cr–16Ni model austenitic alloy. *J. Nucl. Mater.*, **2007**, *367-370*, 930-934.

[62]   Druzhkov, A.P.; Arbuzov, V.A.; Perminov, D.A. Accumulation and annealing of radiation defects in deformed alloys. *Fiz. Metal. i Metalloved.*, **2002**, *94*, N 1, 75-79.

[63]   Dunlop, A.; Lesueur, D.; Lorenzelli, N.; Audouard, A.; Dimitrov, C.; Ramillon, J.M.; Thomé, L. Search for damage and/or disordering effects due to intense electronic excitation in crystalline metallic alloys irradiated by high-energy heavy ions. *J. Phys. Condens. Matter*, **1990**, *2*, 1733-1741.

[64]   Bakay, A.S.; Sergeeva, G.G.; Fateev, M.P. Effect of gamma radiation on the electronic structure of metals. *VANT, Fiz. Rad. Povr. i Rad. Materialoved.*, **1983**, *3/26*, 32-32.

[65]   Schüle, W. On the validity of the one- and two-interstitial model. *Rad. Eff. Lett.*, **1980**, *50*, 93-96.

[66]   Massey H.S.; Burhop, E.H.S.; Gibody, H.B. *Electronic and Ionic Impact Phenomena*, 2nd ed.; Clarendon Press: Oxford, **1969**.

[67]   Zinkle, S.J.; Kulcinski, G.L. 14-MeV neutron irradiation of copper alloys. *J. Nucl. Mater.*, **1984**, *122*, 449-454.

[68]   Sattonnay, G.; Ma, F.; Dimitrov, C.; Dimitrov, O. Radition-induced defects in electron-irradiated γ-TiAl compounds: the effect of composition. *J. Phys.: Condens. Mater.*, **1997**, *9*, 5527-5541.

[69]   Dimitrov, C.; Sitaud, B.; Zhang, X.; Dimitrov, O.; Dedek, U.; Dworschak, F. Radiation-induced defects in solid solutions and intermetallic compounds based on the Ni-Al system: I. Low-temperature electron-irradiation damage. *J. Phys.: Condens. Matter.*, **1992**, *4*, 10199-10210.

[70]   Pravdjuk, N.F.; Ivanov, A.N.; Dubrovin, K.P. Effect of reactor irradiation on materials of thermo-couples. *Atom energy*, **1968**, *25*, 233-235.

[71]   Adamenko, A.A.; Dekhtyar, I.Ya.; Shalaev, A.M. Influence of irradiation on absolute thermal E.M.F. of metals and alloys. *Fiz. Metal. i Metalloved.*, **1972**, *34*, 464-469.

[72]   Yesin, I.A.; Rudnev, I.A. Influence of radiation defects during ion irradiation on critical current of Nb₃Sn superconducting films. *Fiz. Metal. i Metalloved.*, **1988**, *66*, 486-489.

[73]   Karkin, A.Ye.; Mirmelshteyn, A.V.; Arkhipov, V.E.; Goshchitskiy, B.N. Heat capacity of Nb₃Sn and V₂Zr compounds irradiated by large fluence of fast neutrons. *Fiz. Metal. i Metalloved.*, **1987**, *63*, 893-899.

[74]   Sweedler, A.R.; Cox, D.E. Superconductivity and atomic ordering in neutron-irradiated Nb₃Al. *Phys. Rev. B*, **1975**, *12*, 147-156.

[75]   Schilling, W.; Sonnenberg, K. Recovery of irradiated and quenched metals. *J. Phys. F*, **1973**, *3*, 322-350.

[76]   Nikolaev, A.L. Stage I of recovery in 5 Mev electron-irradiated iron and iron-chromium alloys: the effect of small cascades, migration of di-interstitial and mixed dumbbells. *J. Phys. Condens. Matter*, **1999**, *11*, 8633-8644.

[77]   Swanson, M.L.; Maury, F. The location of displaced impurity atoms in irradiated aluminum crystals by backscattering. *Canad. J. Phys.*, **1975**, *53*, 1117-1140.

[78]   Habtetsion, S.; Blythe, H.J.; Dworschak, F.; Dedek, U. Resistivity recovery in cobalt following electron irradiation at 8 K. *J. Phys.: Condens. Matter*, **1989**, *1*, 9519-9532.

[79]   Bois, P.; Beuneu, F. Annealing of defects created by electron irradiation in bismuth. *J. Phys. Condens. Matter*, **1989**, *1*, 4535-4542.

[80]   Pfeiler, W.; Poerschke, Defect recovery and short-range order in some concentrated alloys after electron irradiation at 77 K. R. *J. Phys. F*, **1987**, *17*, 1043-1050.

[81]   Ehrhard, P.; Schlagheck, U. Investigation of Frenkel defects in electron irradiated copper by Huang scattering of x rays: II. Thermal annealing. *J. Phys. F*, **1974**, *4*, 1589-1598.

[82]   Kozlov, A. V. Main mechanisms of influence of structure changes in austenitic steel during low-temperature neutron irradiation on its physics-mechanical properties. *Fiz. Metal. i Metalloved.*, **1966**, *81*(3), 97-106.

[83]   Osetsky, Yu.N.; Bacon, D.J.; Singh, B. N.; Wirth, B Atomistic study of the generation, interaction, accumuiation and annihilation of cascade-induced defect clusters. *J. Nucl. Mater.*, **2002**, *307-311*, 852-861.

[84]   Soisson, F. Kinetic Monte Carlo simulations of radiation induced segregation and precipitation. *J. Nucl. Mater.*, **2006**, *349*, 235-250.

[85]   Dudarev, S.L.; Derlet, P.M. Molecular dynamics modelling of radiation defects in ferromagnetic α-iron. *J. Nucl. Mater.*, **2007**, *367-370*, 251-256.

[86]   Alig, R. C.; Bloom, S. Electron-hole pair creation energies in semiconductors. *Phys. Rev. Lett.*, **1975**, *35*, 1522-1525.

[87]   Lushchik, Ch. B. *Creation of Frenkel Pairs by Excitons in Alkali Halides*. In: Physics of Radiation Effects in Crystals. Ed. R.A. Johnson, A.N. Orlov. Elsevier Science Publ., Amsterdam, **1986**; 473-525.

[88]   Lushchik, N.E.; Vasil'chenko, E.A.; Soovik, Kh.A. In: *Luminescence (Crystal-Phosphorus)*, Proceedings of the 21th Vsesoyuz. Sov. on the Luminescence, Stavropol, USSR, **1973**; p. 108.

[89]   Hersh, H.H. Proposed excitonic mechanism of color center formation in alkali-halides. *Phys. Rev.*, **1966**, *148*, 928-932.

[90]   Pooley, D. F centre production in alkali halides by radiationless electron hole recombination. *Sol. St. Comm.*, **1965**, *3*, 241-243.

[91]   Schwartz, K.; Wirth, G.; Trautmann, C. Ion-induced formation of colloids in LiF at 15 K. *Phys. Rev. B*, **1997**, *56*, 10711-10714.

[92]   Tanimura, K.; Itoh, N. Selective non-radiative transitions at excited states of the self-trapped exciton in alkali halides. *J. Phys. Chem. Solids*, **1984**, *45*, 323-340.

[93]   Lushchik, Ch.B; Savikhin, F.A.; Makhov, V.N.; Ryabukhin, O.V.; Ivanov, V.Yu.; Kruzhalov, A.V.; Neshov, F.G. Recombination-assisted creation of cation excitons and cross-luminescence quenching in CsCl crystals at high excitation densities. *Phys. Sol. State*, **2000**, *42*, 1052-1057.

[94]   Luschik, Ch.B.; Vitol, I.K.; Elango, M.A. Exciton mechanism of F-center formation in defect-free parts of ionic crystals. *Fiz. Tverd. Tela*, **1968**, *10*, 2753-2759.

[95]   Varley, Y. A mechanism for the displacement of ions in an ionic lattice. *Nature*. **1954**, *174*, 886-887.

[96]   Seitz, F. Color centers in alkali halide crystals. *Rev. Modern Phys.* **1954**, *26*, 7-34.

[97]   Bhuniya, R.C.; Rao, S.E. Colour centres and luminescence in KCl:Zn crystals. *Phys. Stat. Sol. (b)*, **1985**, *131*, 339-348.

[98]   Lushchik, A.; Kudrjavtseva, I; Lushchik, Ch.; Vasil'chenko, E. Creation of stable Frenkel defects by vacuum uv radiation in KBr crystals under conditions of multiplication of electronic excitations. *Phys. Rev.*, **1995**, *B 52*, 10069-10072.

[99]   Baimakhanov, A.; Yigi, H.R.-V.; Lushchik, A. Ch. Homo- and heterogeneous radiation defect distribution in KCl. *Fiz. Tverd. Tela*, **1987**, *29*, 1356-1363.

[100]   Schwartz, K. Electronic excitations and defect creation in LiF crystals. *Nucl. Instr. Meth.*, **1996**, *B 107*, 128-132.

[101]   Perez, A.; Balanzar, E.; Dural, J. Experimental study of point-defect creation in high-energy heavy-ion tracks. *Phys. Rev.*, **1990**, *B 41*, 3943-3950.

[102]   Sobolev, A.B.; Varaksin, A.N. Shell model calculation of lattice deformation in the vicinity of an F-center in NaCl. *Fiz. Tverd. Tela*, **1994**, *36*, 275-283.

[103]   Kerr, R.K.; Schwerdtfeger, C.F. Role of interstitial aggregation on the F-coloring curve of alkali halides. *J. Phys. Chem. Solids*, **1971**, *32*, 2009-2013.

[104]   Melik-Gayakazjan, I.Ya.; Deryabin, P.E.; Turgumbaev K.T. In: *Luminescence (Crystal-Phosphorus)*, Proceedings of the 21th Vsesoyuz. Sov. on the Luminescence, Stavropol, USSR, 1973; p. 184.

[105]   *Physics of Color Center*. Ed. by W.B. Fowler; Academic Press: New York and London, **1968**.

[106]   Ivakhnenko, P.S.; Vilitkevich, A.I. In: *Physics of the Solid State*; To question on color centers and thermal luminescence of phosphor-crystal, KGPI: Kharkov, **1975**; pp. 12-19.

[107]   Shuraleva, E.I.; Ivakhnenko, P.S. In: *Physics of the Solid State*; Analysis of accumulation kinetics of F centers in AHC with Eu^{++}, KGPI: Kharkov, **1975**; pp. 40-46.

[108]   Bichevin, V.; Kaambre, H.; Nakonechnyi, S. Temperature dependence of the excimer laser coloration of KBr. *Phys. Stat. Sol. (a)*, **1998**, *168*, 55-61.

[109]   Nakonechnyi, S.; Kärner, T.; Lushchik, A.; Lushchik, Ch.; Babin, V.; Feldbach, E.; Kudryavtseva, I.; Liblik, P.; Pung, L.; Vasil'chenko, E. Low-temperature excitonic, electron-hole and interstitial-vacancy processes in LiF single crystals. *J. Phys.: Condens. Matter*, **2006**, *18*, 379-394.

[110]   Bosi, L.; Nimis M. On the decay properties of the M center in LiF crystals. *Phys. Stat. Sol. (b)*, **1989**, *152*, 67-71.

[111]   Mikhailov, M.M.; Ardyshev, V.M.; Belyakov, M.V. Oscillator strength of electron-type color centers in KCl single crystals irradiated with electrons and protons. *Phys. Sol. Stat.*, **2002**, *44*, 274-277.

[112]   Alukher, E.D.; Lusis, L.Yu.; Chernov, C.A. *Electron Excitation and Radioluminescence of Alkali Halide Crystals*; "Zinatne": Riga, **1979**.

[113]   Akilbekov, A.; Elango, A. Low-temperature pair associates of H centers in KBr. *Phys. State. Sol. (b)*, **1984**, *122*, 715-723.

[114]   Akilbekov, A.; Dauletbekova, A.; Elango, A. Photo- and thermochemical reactions with participation of Br$_3^-$ centers in X-reyed KBr. *Phys. Stat. Sol. (b)*, **1985**, *127*, 493-501.

[115]   Rzepka, E.; Lefrant, S.; Taurel, L. Stability of V-type defects in Ca^{2+} and Tl$^+$ potassium iodide crystals. *Rad. Eff.*, **1986**, *98*, 301-304.

[116]   Castro, M.J. Volume expansion in heavily gamma irradiated NaCl and LiF single crystals. *J. Phys.: Condens. Matter*, **2008**, *20*, 095211.

[117]   Dubinko, V.I.; Vainshtein, D.I.; Den Hartog, H.W. Effect of radiation-induced emission of schottky defects on the formation of colloids in alkali halides. *Rad. Eff.*, **2003**, *158*, 705-719.

[118]   Frugoli, P.A.; Pimentel, C.A. Point defect aggregates in γ-irradiated LiF single crystals. *Phys. Stat. Sol. (b)*, **1983**, *117*, 549-556.

[119]   Shapurko, A.V.; Gromov, L.A.; Kukushkin, SA.; Shtanko, V.I. Diffusive growth of vacancy pores in irradiated CsBr. *Fiz. Tverd. Tela*, **1988**, *30*, 3493-3496.

[120]   Itoh, N.; Tanimura, K. Radiation effects in ionic solids. *Rad. Eff.*, **1986**, *98*, 269-287.

[121]   Hobbs, L.W.; Hughes, A.E.; Pooley, D. A study of interstitial clusters in irradiated alkali halides using direct electron microscopy. *Proc. Roy. Soc. Lond. A*, **1973**, *332*, 167-185.

[122]   Andronikashvili, E.L.; Galustashvili, M.V.; Driyaev, D.G.; Saralidze, Z.K. The nature of new dislocations in neutron-irradiated LiF crystals. *Fiz. Tverd. Tela*, **1987**, *29*, 130-135.

[123]   Gektin, A.V. Effect of radiation point and linear defects on hardening of NaCl type crystals. *Fiz. Tverd. Tela*, **1985**, *27*, 3254-3256.

[124]   Levy, P.V. Radiation damage studies on non-metals utilizing measurements made during irradiation. *J. Phys. Chem. Solids*, **1991**, *52*, 319-349.

[125]   Gotlib, V.I.; Khristapson, Ya.Zh.; Schwartz, K.K.; Ekhmanis, Yu.A. In: *Radiation Physics VII. Migration of Energy and the Defects in Alkali Halide Crystals*; Colloidal centers and the process of radiolysis in alkali halide crystals, Ed.; "Zinatne": Riga, **1973**; pp. 143-196.

[126]   Zefirova, V.L.; Kolontsova, E.V.; Telegina, L.V. Defects in electron-irradiated LiF single crystals at various stages of irradiation and annealing. *Kristallografiya*, **1975**, *20*, 588-591.

[127]   Soppe, W.J. Computer simulation of radiation damage in NaCl using a kinetic rate reaction model. *J. Phys.: Condens. Matter*, **1993**, *5*, 3519-3540.

[128]   Van der Kolk, G.J.; Post, K.; Van Veen, A.; Pleiter, F.; De Hosson, J.Th.M. Interaction of vacancies with implanted metal atoms in tungsten observed by means of thermal helium desorption spectrometry and perturbed angular correlation measurement. *Rad. Eff.*, **1985**, *84*, 131-158.

[129] Ei Sayed, H.; Kovács, I. Binding energy between a vacancy and Fe atom in dilute Al-Fe alloys. *Phys. Stat. Sol. (a)*, **1975**, *27*, K35-37.

[130] Arbuzov, V.L.; Danilov, S.Ye.; Klotsman, S.M.; Petrusenko, Yu.T.; Sleptsov, A.N. Radiation defect annealing in region of III stage in Nickel-Yttrium alloy. *Fiz. Metal. i Metalloved.*, **1990**, *90*, N 12, 156-157.

[131] Hackett, M.J.; Najafabadi, R.; Was, G.S. Modeling solute-vacancy trapping at oversized solutes and its effect on radiation-induced segregation in Fe–Cr–Ni alloys. *J. Nucl. Mater.*, **2009**, *389*, 279-287.

[132] Maury, F.; Lucasson, P.; Lucasson, A.; Vajda, P.; Balanzat, E.; Beretz, D; Halbwachs, M.; Hillairet, J. Single and multiple trapping of radiation-induced defects in AgZn alloys. *Rad. Eff.*, **1984**, *82*, 141-153.

[133] Dimitrov, C.; Dimitrov, O.; Dvorschak, F. The interaction of self interstitial with undersized solute atoms in electron-irradiated aluminum. *J. Phys. F*, **1978**, *8*, 1031-1052.

[134] Mansel, W.; Vogl, G. Fast neutron radiation damage in aluminium studied by interstitial trapping at Co Mössbauer atoms. *J. Phys. F*, **1977**, *7*, 253-271.

[135] Lam, N.Q.; Dagens, L.; Doan, N.V. Molecular dynamics study of interstitial-solute interactions in irradiated alloys: III. Configurations and binding energies of interstitial-solute complexes in Al-Cu, Al-Ag and Al-Au alloys. *J. Phys. F*, **1983**, *13*, 1369-1377.

[136] Lam, N.Q.; Dagens, L.; Doan, N.V. Molecular dynamics study of interstitial-solute interactions in irradiated alloys: II. Configurations and binding energies of interstitial-solute complexes in Al-Be, Al-Ca, Al-K, Al-Li and Al-Mg alloys. *J. Phys. F*, **1981**, *11*, 2231-2245.

[137] Lam, N.Q.; Doan, N.V.; Adda, Y. Molecular dynamics study of interstitial-solute interactions in irradiated alloys: I. Configurations, binding and induced migration energies of mixed dumbbells in Al-Zn alloys. *J. Phys. F*, **1980**, *10*, 2359-2373.

[138] Abe, H.; Kuramoto, E. Interaction of solutes with irradiation-induced defects of electron-irradiated dilute iron alloys. *J. Nucl. Mater.*, **1999**, *271-272*, 209-213.

[139] Rehn, L.E.; Robrock, K-H.; Jacques, H. Interstitial-solute complexes in an irradiated Al-Fe alloy. *J. Phys. F*, **1978**, *8*, 1835-1844.

[140] Dworschak, F.; Monsau, Th.; Wollenberger, H. Impurity-interstitial interaction in electron-irradiated aluminium. *J. Phys. F*, **1976**, *6*, 2207-2218.

[141] Maury, F.; Lucasson, P.; Lucasson, A.; Vajda, P.; Balanzat, E.; Beretz, D; Halbwachs, M.; Hillairet, J. Single and multiple trapping of radiation-induced defects in AgZn alloys. *Rad. Eff.*, **1984**, *82*, 141-153.

[142] Wollenberger, H. Interaction of self-interstitials with solutes. *J. Nucl. Mater.*, 1978, *69-70*, 362-371.

[143] Maury, F.; Lucasson, A.; Lucasson, P.; Moser, P.; Faudot, F. Interstitial migration in irradiated iron and iron-based dilute alloys: I. Interstitial trapping and detrapping in FeMo, FeV and FeTi dilute alloys. *J. Phys.: Condens. Matter*, **1990**, *2*, 9269-9290.

[144] Maury, F.; Lucasson, A.; Lucasson, P.; Moser, P.; Faudot, F. Interstitial migration in irradiated iron and iron-based dilute alloys: II. Interstitial migration and solute transport in FeNi, FeMn and FeCu dilute alloys. *J. Phys.: Condens. Matter*, **1990**, *2*, 9291-9307.

[145] Arbuzov, V.L.; Danilov, S.E.; Davletshin, A.E.; Zuev, Yu.L.; Andryushkin, V.V. Interaction of radiation defects with the deuterium and tritium atoms in nickel. *Fiz. Metal. i Metalloved.*, **1988**, *65*, N 5, 79-83.

[146] Swanson, M.L.; Howe, L.M.; Quenneville, A.F. Irradiation-induced displacement of Ag atoms from lattice sites in an Al-0.2Mg-0.1Ag crystal. *J. Phys. F*, **1976**, *6*, 1629-1637.

[147] Bochkarev, V.V.; Danilov, V.P.; Murina, T.M.; Prokhorov, A.M. Optical Formation of A° (1) Centers in KBr:Tl, KCl:Tl, and KCl:In Crystals. *Phys. Stat. Sol. (b)*, **1992**, *173*, K43-46.

[148] Tsuboi, T.; Kamewari, J. W. Magnetic circular dichroism spectra of the Ag-center in KCl crystals. *Phys. Stat. Sol. (b)*, **1994**, *185*, K43-46.

[149] Kleeman, W. Electron-lattice interaction of Ag⁻ and Cu⁻ centers in alkali halides. *Z. Phys.*, **1971**, *249*, 145-167.

[150] Lüty, F.; Costa Ribeiro, S.; Mascarenhas, S.; Sverzut, V. Photoelastic measurements of the volume expansion by the U→F and U→α transformation in KBr. *Phys. Rev.*, **1968**, *168*, 1080-1086.

[151] Egranov, A.V.; Chernyago, B.P. Atomic hydrogen at anion sites in LiF-H⁻ and NaF-H⁻ crystals. *Phys. Stat. Sol. (b)*, **1995**, *188*, 615-621.

[152] Akhvlediani, Z.G.; Akhvlediani, I.G.; Kalabegishvili, T.L. Effect of combined mechanical load and irradiation on hydrogen stabilization in LiF:OH. *Phys. Stat. Sol. (b)*, **1983**, *119*, 503-506.

[153] Egranov, A.V.; Nepomnyachikh, A.I. Magnesium colour centers in NaF. *Phys. Stat. Sol. (b)*, **1984**, *122*, 249-254.

[154] Bosi, L.; Nimis, M. On analysis of the classical arguments concerning forecasts for the Z_1 centre formation in alkali halide. *Phys. Stat. Sol. (b)*, **1985**, *131*, K111-116.

[155] Stoicescu, Gh.; Nistor, S.V.; Mateescu, C.D. Aggregation of bismuth in NaCl crystals. *Phys. Stat. Sol. (b)*, **1989**, *156*, 411-418.

[156] Nierzewski, K.D.; Macalik, B.; Opyrchal, H. Effect of low temperature γ-irradiation on I-V dipoles in KCl-Eu²⁺ crystals. *Phys. State. Sol. (b)*, **1986**, *133*, K91-93.

[157] Ivakhnenko, P.S.; Parfianovich, I.A.; Shuraleva, E.I. The formation of F centers and electronic recombination luminescence in alkali halide crystals doped with Eu²⁺. *Izv. AN SSSR, Phys.*, **1969**, *33*, 844-847.

[158] Kao, K.J.; Perlman, M.M. X-ray effects on cation impurity-vacancy pairs in KCl-Eu²⁺. *Phys. Rev. B*, **1979**, *19*, 1193-1202.

[159] Bhunita, R.C.; Rao, S.E. Color centers and luminescence in KCl-Ni. *Phys. Stat. Sol. (b)*, **1982**, *114*, 561-568.

[160] Kawai, T.; Yamano, A. Creation of CsAu nanoclusters by UV-light irradiation on CsBr-Au⁻ crystals. *Phys. Stat. Sol. (b)*, **2006**, *246*, 488-493.

[161] Panova, A.N.; Kudin, A.M.; Dolgopolova, A.V. Thermal stability of electron and hole activator colour centers in NaI-Tl crystals. *Opt. i Spektr.*, **1987**, *63*, 444-445.

[162] Goldberg, L.S. Luminescence from ICl⁻ V_K-center-electron recombination and localized exciton decay in KCl:I. *Phys. Rev.*, **1968**, *168*, 989-991.

[163] Saidoh, M.; Hoshi, J.; Itoh, N. Temperature dependence of the dynamic interstitial interactions in potassium bromide. *Solid State Comm.*, **1973**, *13*, 431-433.

[164] Korepanov, V.I.; Kuznetsov, M.F.; Malyshev, A.A.; Strezh, V.V. H centers in AHC with heavy homologous anion impurity. *Fiz. Tverd. Tela*, **1990**, *32*, 1317-1322.

[165] Salomatov, V.N.; Parfianovich, I.A. The F_H-centre energy structure in KCl and KBr. *Phys. Stat. Sol. (b)*, **1984**, *121*, 401-406.

[166] Bosi, L.; Nimis, M. On the criterions for F_A(II) center formation in alkali halide crystals. *Phys. Stat. Sol. (b)*, **1989**, *156*, K5-10.

[167] Ahlers, F.J.; Baranov, P.G.; Romanov, N.G.; Spaeth, J.-M. ODMP of silver centers in KCl. *Fiz. Tverd. Tela*, **1988**, *30*, 427-432.

[168] Zazubovich, S.; Usarov, A.; Egemberdiev, Zh. Photothermal creation of Cu°Va⁺ centres and mobility of anion vacancies in X-irradiated KCl:Cu crystals. *Phys. Stat. Sol. (b)*, **1983**, *118*, 789-798.

[169] Radhakrishna, S.; Chowdari, B.V.R. Z centers in impurity-doped alkali halides. *Phys. Stat. Sol. (a)*, **1972**, *14*, 11-39.

[170] Yazici, A.N.; Öztürk, Z. The defect structure of new glow peak generated from heavily annealed LiF:Mg, Ti. *Phys. Stat. Sol. (b)*, **1998**, *209*, 195-204.

[171]   Egemberdiev, Zh.; Ismailov, K.; Usarov, A.; Zazubovich, S.; Jaanson, N. Luminescent Associates of $Pb^{2+}V_c^-$ Dipoles with Interstitial Iodine Atoms in $KI:PbI_2$ Crystals. *Phys. Stat. Sol. (b)*, **1991**, *163*, 183-190.

[172]   Egemberdiev, Zh.; Usarov, A.; Zazubovich, S. Luminescence of Lead Ions Associated with Interstitials and Vacancies in Alkali Halides. *Phys. Stat. Sol. (b)*, **1991**, *164*, 195-206.

[173]   Ikeya, M. Neutral manganese centers at the anion site in NaCl. *Phys. Stat. Sol. (b)*, **1972**, *51*, 407-414.

[174]   Aceves, R.; Barboza Flores, M.; Nagirnyi, V.; Perez Salas, R.; Usarov, A.; Zazubovich, S. Luminescent associates of indium ions with interstitials and vacancies in an x-irradiated KCl:In crystal. *Phys. Stat. Sol. (b)*, **1996**, *195*, 439-450.

[175]   Van Puymbroeck, W.; Schoemaker, D. Electron-spin-resonance study of complex interstitial halogen H_D-type defects in KCl doped with divalent cations. *Phys. Rev. B*, **1981**, *23*, 1670-1684.

[176]   Grasa Molina, M.I.; Pawlik, Th.; Spaeth, J.-M. Investigation of radiation-induced defects in $NaBr:Sr^{2+}$. *Phys. Stat. Sol. (b)*, **1997**, *202*, 993-998.

[177]   Pawlik, Th.; Nistor, S.V.; Spaeth, J-M. Electron-nuclear double-resonance study of a substitutional Fe^{3+} defect complex in x-irradiated NaCl crystals. *J. Phys.: Condens. Matter*, **1997**, *9*, 7631-7642.

[178]   Kurobori, T.; Nebel, A.; Beigang, R.; Welling, H. Performance characteristics of an NaCl:OH$^-$ $(F^+_2)_H$ color center laser pumped by mode-locked $Nd:LiYF_4$ laser. *Opt. Commun.*, **1989**, *73*, 365-369.

[179]   Salomatov, V.N.; Yur'eva, T.G. Energy and spatial structure of $(F_2^+)_H$-centers in NaCl, KCl and KBr. *Fiz. Tverd. Tela*, **1991**, *33*, 1801-1804.

[180]   Alekseev, P.D.; Baranov, G.I.; Kurakina, E.P.; Maltsev, K.A. Formation of hydrogen bonding in doped OH-group alkali halide crystals by the action of γ-radiation. *Phys. Stat. Sol. (b)*, **1983**, *120*, K119-121.

[181]   Alekseev, P.D.; Belyaeva, V.K.; Marov, I.N. Hydrogen-bonded centers in the γ-irradiated LiF host lattice and their manifestation in IR and ESR spectra. *Fiz. Tverd. Tela*, **1988**, *30*, 308-311.

[182]   Egranov, A.V.; Otroshok, V.V.; Chernyago, B.P. Atomic Hydrogen Centres in LiF-H$^-$ and LiF-H$^-$, Mg^{2+} Crystals. *Phys. Stat. Sol. (b)*, **1991**, *167*, 451-458.

[183]   Grigorjev, V.A.; Lyapidevskii, V.K.; Obodovskii, I.M. The effect of electric field on thermoluminosity of alkali halide scintillators. *Fiz. Tverd. Tela*, **1969**, *4*, 1042-1045.

[184]   Delgado, L. Radiation induced aggregation and thermoluminescence of copper centers in KCl. *Rad. Eff.*, **1983**, *73*, 45-51.

[185]   Murti, Y.V.G.S.; Murthy, K.R.N. Thermoluminescence of alkali halides irradiated at 80 K. *J. Phys. C*, **1974**, *7*, 1918-1928.

[186]   Kuketaev, T. Activator trapping centers in copper doped alkali halide crystals. *Tr. IFA AN ESSR*, **1969**, *35*, 196-201.

[187]   Smolskaya, L.P.Recombination luminescence of KI-Cu phosphorus. *Izv. AN SSSR, Phys.*, **1969**, *33*, 1020-1022.

[188]   Nistor, S.V.; Ursie, I.; Goovaerts, E.; Schoemaker, D. Structural, optical and production properties of $Tl^0(1)$ laseractive center in NaCl. *Rev. Roum. Phys.*, **1986**, *31*, 865-879.

[189]   Reddy, K.N.; Babu, V.H. The study of Z_2-centres in NaCl crystals doped with samarium. *Phys. State. Sol. (a)*, **1982**, *74*, K127-130.

[190]   Reddy, K.N.; Rao, M.L.; Babu, V.H. Thermoluminescence and optical absorption studies of Z_1-centres in NaCl crystals doped with samarium. *Phys. State. Sol. (a)*, **1982**, *74*, K127-130.

[191]   Reddy, N.K.; Rao, S.U.V.; Babu, B. A study of Z_1-centers in γ-irradiated $KCl:Ba^{2+}$ crystals. *Cryst. Res. Technol.*, **1983**, *18*, 1401-1405.

[192]   Reddy, K.N.; Ahmed, Md.I.; Pandaraiah, N.; Rao, S.U.V.; Babu, V.H. A thermoluminescence stady of Z_2-centres in terbium-doped NaCl crystals. *Cryst. Res. Technol.*, **1983**, *18*, 1155-1160.

[193]   Sridaran, P.; Gartia, R.K.; Bhuniya, R.C.; Ratnam, V.V. Thermoluminescence of Z_2 centers in X-irradiated Ca-doped KCl crystals. *Phys. Stat. Sol. (a)*, **1981**, *64*, 127-131.

[194]   Reddy, N.K.; Rao, S.U.V.; Babu, B. State of dispersion of impurity on formation of Z_1-centres in $NaCl:Sa^{2+}$ crystals. *Cryst. Res. Technol.*, **1983**, *18*, 1291-1298.

[195]   Bhan, S. Thermoluminescence of $NaF:Ca^{2+}$ single crystals. *Phys. Stat. Sol. (b)*, **1982**, *112*, 507-514.

[196]   Sastry, S.B.S.; Nagarajan, S. Thermoluminescence and optical absorption studies rubidium bromide doped with barium. *Phys. Stat. Sol. (b)*, **1983**, *117*, 171-175.

[197]   Veeresham, P.; Reddy, K.N.; Rao, U.V.S.; Babu, V.H. Thermoluminescence and optical absorption studies of KBr crystals doped with Ca^{2+}. *Crystal Res. and Technol.*, **1983**, *18*, 1427-1431.

[198]   Moharil, S.V.; Kamavisdar, V.S.; Deshmukh, B.T. Thermoluminescence of Z_1-centres in NaCl-Ca. *Phys. Stat. Sol. (a)*, **1979**, *55*, K167-171.

[199]   Pode, R.B. Formation of the Z_1 colour centre by mechanical bleaching in γ-irradiated KCl:Ca. *Phys. Stat. Sol. (a)*, **1986**, *96*, K147-149.

[200]   Sastry S.B.S.; Viswanathan V.; Ramasastry C. Lead centres in alkali halides: NaCl, KCl and KBr. *J. Phys. Soc. Jap.*, **1973**, *35*, 508-513.

[201]   Schoemaker, D.; Kolopus, J.L. Pb^{2+} as a hole trap in KCl:ESP and optical absorption of Pb^{3+}. *Sol. Stat. Commun.*, **1970**, *8*, 435-439.

[202]   Dimitrov, C.; Dimitrov, O. Composition dependence of defect properties in electron-irradiated Fe-Cr-Ni solid solutions. *J. Phys. F*, **1984**, *14*, 793-811.

[203]   Nakata, K.; Kato, T.; Masaoka, I. Void formation and precipitation during electron-irradiation in austenitic stainless steels modified with Ti, Zr and V. *J. Nucl. Mater.*, **1987**, *148*, 185-193.

[204]   Kato, T.; Takahashi, H.; Izmiya, M. Grain boundary segregation under electron irradiation in austenitic stainless steels modified with oversized elements. *J. Nucl. Mater.*, **1992**, 189, 167-174.

[205]   Mayer, R.M.; Morris, E.T. Neutron irradiation of dilute aluminium alloys. *J. Nucl. Mater.*, **1977**, *71*, 36-43.

[206]   Takeyama, T.; Ohnuki, S.; Takahashi, H. The effect of precipitation on void formation in copper-iron alloy during electron irradiation. *J. Nucl. Mater.*, **1980**, *89*, 253-262.

[207]   Zee, R.H.; Birtcher, R.C.; MacEwen, S.R. Effects of solute on damage production and recovery in zirconium. *J. Nucl. Mater.*, **1986**, *141-143*, 771-775.

[208]   Wu, Y.C.; Itoh, Y.; Ito, Y. Positron annihilation studies on the interaction between hydrogen and defects in nickel. *Phys. Stat. Sol. (b)*, **1996**, *193*, 307-310.

[209]   Krishan, K. Invited review article ordering of voids and gas bubbles in radiation environments. *Rad. Eff.*, **1982**, *66*, 121-155.

[210]   Leinaker, J. M.; Bloom, E. E.; Stigler, J. O. The effect of minor constituents on swelling in stainless steel. *J. Nucl. Mater.*, **1973/1974**, *49*, 57-66.

[211]   Averback, R.S.; Ehrhart, P. Diffuse x-ray scattering studies of defect reaction in electron-irradiated dilute nickel alloys. I: Ni-Si. *J. Phys. F*, **1984**, *14*, 1347-1363.

[212]   Alyab'yev, V.M; Vologin, V.G.; Dubinin, S.F.; Lapin, S.S.; Parkhomenko, V.D.; Sagaradze, V.V. Neutron diffraction and electron-microscope studies of precipitation processes and radiation-stimulated ageing of Cr-Ni-Ti austenitic steels. *Fiz. Metal. i Metalloved.*, **1990**, *69*, N 8, 142-148.

[213]   Wiffen, F.W.; Maziasz, P.J. The influence of neutron irradiation at 55°C on the properties of austenitic stanless steels. *J. Nucl. Mater.*, **1981**, *103-104*, 821-826.

[214]   Platov, Yu.M.; Votinov, S.N.; Drits, M.E.; Ivanov, L.I.; Kalinin, V.G.; Smirnov, A.V.; Toropova, L.S.; Shamardin, V.K. Investigation of mechanical properties of alloys on the basis of aluminum after neutron irradiation. *Fiz. Chim. Obrab. Mater.*, **1981**, N1, 53-55.

[215]   Repnikova, Ye.A.; Malinenko, I.A.; Chudinova, S.A.; Toropova, L.S.; Ustinovshchikov, V.M. Electron irradiation influence on precipitation of Al-Mg-Sc alloy. *Fiz. Metal. i Metalloved.*, **1984**, *57*, 531-534.

[216]   Druzhkov, A.P.; Perminov, D.A.; Arbuzov, V.L. Effects of intermetallic nanoparticles on the evolution of vacancy defects in electron-irradiated Fe–Ni–Al material. *J. Phys. Condens. Matter*, **2006**, *18*, 365-377.

[217]   Arbuzov, V.L.; Druzhkov, A.P.; Danilov, S.E. Effects of phosphorus on defects accumulation and annealing in electron-irradiated Fe–Ni austenitic alloys. *J. Nucl. Mater.*, **2001**, *295*, 273-280.

[218]   Bekeshev, A.Z.; Vasil'chenko, E.A.; Dauletbekova, A.K.; Shunkeev, K.Sh.; Elango, A.A. Radiation-stimulated output of impurities in interstitial sites in crystals of KBr-Li and KCl-Li. *Fiz. Tverd. Tela*, **1996**, *38*, 769-778.

[219]   Giuliani, G. Colour centres production in sodium doped KBr at 80°K. *Sol. State Commun.*, **1969**, *7*, 79-82.

[220]   Still, P.B.; Pooley, D. F-centre production in mixed alkali halide crystals as evidence for the importance of a replacement collision sequence in F-centre production. *Phys. Stat. Sol.*, **1969**, *32*, K147-150.

[221]   Hirai, M. The yield of color center formation in KBr doped with KI. *Sol. State Commun.*, **1972**, *10*, 493-495.

[222]   Nagli, L.E.; Karklinja, M.N. Self-localized excitons in CsI and CsI(Na). *Fiz. Tverd. Tela*, **1989**, *31*, N 12, 160-163.

[223]   Malyshev, A.A.; Yakovlev, V.Yu. Relaxed heteronuclear excitons in KCl:I crystal. *Fiz. Tverd. Tela*, **1982**, *24*, 2296-2299.

[224]   Ikeya, M.; Itoh, N.; Okada, T.; Suita, T. Study of the enhancement of X-ray coloration of NaCl by divalent impurities. *J. Phys. Soc. Jap.*, **1966**, *21*, 1304-1309.

[225]   Sonder, E.; Bassignani, G.; Camagni, P. Impurities and secondary reactions in radiation defect production at liquid-nitrogen temperature in alkali halides. *Phys. Rev.*, **1969**, *180*, 882-889.

[226]   Opyrchal, H.; Nierzewski, K.D.; Macalik, B.; Mladenova, M. γ-ray induced colaration of KCl-Eu^{2+} crystals. *Phys. Stat. Sol. (b)*, **1986**, *135*, 141-148.

[227]   Ikeya, M.; Itoh, N.; Okada, T.; Suita, T. Study of the enhancement of X-ray coloration of NaCl by divalent impurities. *J. Phys. Soc. Jap.*, **1966**, *21*, 1304-1309.

[228]   Chowdari, B.V.R.; Itoh N. X-Ray coloration of Eu^{2+}-doped KCl. *Phys. Stat. Sol. (b)*, **1971**, *46*, 549-557.

[229]   Medrano, C.P.; Ramos, S.B.; Hernandez, J.A.; Murrieta, H.S.; Zaldo, C.; Rubio, J.O. Influence of radiation intensity and lead concentration in the room-temperature coloring of KBr. *Phys. Rev. B*, **1985**, *32*, 6837-6844.

[230]   Rubio, J.O.; Aguilar, M.G.; Lopez, F.J.; Galan, M.; Garsia-Sole, J.; Murrieta, H.S. Effects of X-irradiation in europium-doped NaCl. *J. Phys. C*, **1982**, *15*, 6113-6128.

[231]   Rubio J.O.; Flores, M.C.; Murrieta, H.S.; Hernandez, J.A. Influence of concentration and aggregation-precipitation state of divalent europium in the room-temperature coloring of KCl. *Phys. Rev. B*, **1982**, *26*, 2199-2207.

[232]   Ramos, S.B.; Cordero-Borboa, A.; Murrieta, H.S.; Rubio, J.O. First stage coloration of alkali halides doped with M^{2+} impurities that change their valence state by irradiation as a function of X-irradiation dose rate. *Sol. State Commun.*, **1985**, *56*, 435-438.

[233]   Sanchez, C.; Lerma, I.S.; Jaque, F.; Agulló-López, F. Influence of the aggregation state of Ca^{2+} ions on the colouring and hardening behavior of NaCl:Ca^{2+}. *Crystal Lattice Defects*, **1976**, *6*, 227-232.

[234]   Grigoruk, L.V.; Melik-Gaikazjan, I.Ya. Impurity distribution and formation of colour centres under the action of X-rays in NaCl-Mn, NaCl-Cd, NaCl-Pb. *Opt. i spektr.*, **1963**, *15*, 394-399.

[235]   Gladyshev, G.E. Effect of lead impurities on the radiation resistance of potassium chloride. *Phys Sol. Stat.*, **2006**, *48*, 1893-1895.

[236]   Pascual, J.L.; Agulló-López, F. Influence of concentration and aggregation state of lead on the room-temperature coloring of NaCl and KCl. *Crystal Lattice Defects*, **1977**, *7*, 161-175.

[237]   López, F.J.; Cabrera, J.M.; Agulló-López, F. Radiation-induced colouring in NaCl:Mn^{2+}. *J. Phys. C*, **1979**, *12*, 1221-1238.

[238]   Butterworth, J.S.; Esser, P.D.; Levy, P.W. Formation of color centers in KCl:Tl by gamma irradiation at 21 °C. *Phys. Rev. B*, **1970**, 3340-3353.

[239]   Gavrilov, V.V.; Deich, R.G.; Dyachenko, S.V.; Nagli, L.E.; Pirogov, F.V. On mechanisms of defect formation in activated alkali halide crystals. *Fiz. Tverd. Tela*, **1987**, *29*, 1904-1906.

[240]   Vale, G. The peculiarities of colour centre production in doped alkali halides. *Phys. Stat. Sol. (b)*, **1996**, *197*, 293-297.

[241]   Dauletbekova, A.K.; Gindina, R.I.; Elango, A.A. Aggregative bromine centers in irradiated KBr crystals with impurities. *Opt. i Spektr.*, **1982**, *53*, 548-549.

[242]   Melik-Gaikazjan, I.Ja.; Rostchina, L.I.; Ignatyeva M.I. F-center accumulation in KCl crystals with sulfur impurity. *Fiz. Tverd. Tela*, **1965**, *7*, 3465-3467.

[243]   Tsal, N.A.; Dragan, O.P.; Didyk, R.I. In: *Luminescence (Crystal-Phosphorus)*, Proceedings of the 21th Vsesoyuz. Sov. on the Luminescence, Stavropol, USSR, **1973**; p. 196.

[244]   Didyk, R.I.; Tsal, N.A. F-center formation in NaCl crystals with oxygen impuritiy. *Fiz. Tverd. Tela*, **1972**, *14*, 1840-1843.

[245]   Adda, Y.; Beyeler, M.; Brebec, G. Radiation effects on solid state diffusion. *Thin Solid Films*, **1975**, *25*, 107-156.

[246]   Smirnov, Ye.A.; Mikhin, A.G.; Osetskiy, Yu.N. Radiation-stimulated self-diffusion and point defect characteristics in α-iron. *Fiz. Metal. i Metaloved.*, **1996**, *81*, N 4, 122-125.

[247]   Müller, A.; Naundorf, V.; Macht, M.-P. Material transport parameters for irradiated nickel and austenitic Fe-Cr-Ni alloys. *J. Nucl. Mater.*, **1988**, *155-157*, 1128-1131.

[248]   Bystrov, L.N.; Ivanov, L.I.; Platov, Yu.M. Radiation-enhanced diffusion in metals. *Phys. Stat. Sol. (a)*, **1971**, *7*, 617-627.

[249]   Acker, D.; Beyeler, M.; Brebec, G.; Bendazzoli, M.; Gilbert, J. Effet de l'irradiation aux neutrons sur l'hétérodiffusion á dilution infinie de l'or et du cuivre dans l'aluminium. *J. Nucl. Mater.*, **1974**, *50*, 281-297.

[250]   Naundorf, V. Diffusion in metals and alloys under irradiation. *Int. J. Modern Phys. B*, **1992**, *6*, 2925-2986.

[251]   Wiedersich, H. Evolution of phase microstructure during irradiation. *Rad. Effects*, **1986**, *101*, 21-36.

[252]   Wollenberger, H. Radiation-altered phase stability. *Mater. Sci. Forum*, **1992**, *97-99*, 241-252.

[253]   Shewmon, P., G. *Diffusion in Solids*, McGraw-Hill: New York, **1963**.

[254]   Sizmann, R. The effect of radiation upon diffusion in metals. *J. Nucl. Mater.*, **1968**, *69/70*, 386-412.

[255]   Lam, N.Q.; Rothman, S.J.; Merkle, K.L.; Nowicki, L.J.; Dever, D.J. Effect of proton bombardment on self-diffusion in silver: a preliminary report. *Thin Solid Films*, **1975**, *25*, 157-166.

[256]   Schüle, W.; Kornmann, H. Radiation enhanced diffusion in f.c.c. metals. *Rad. Eff.*, **1980**, *49*, 213-224.

[257]   Kiv, A.; Fuks, D.; Munitz, A.; Zenou, V.; Moiseenko, N. Radiation-stimulated diffusion in Al-Si alloys. *Rad. Eff.*, **2007**, *162*, 59-67.

[258]   Bystrov, L.N.; Ivanov, L.I.; Platov, Yu.M. Non-stationary radiation-enhanced diffusion in metals. Ordering in the cold-worked silver-zinc alloys. *Phys. Stat. Sol. (a)*, **1971**, *8*, 375-381.

[259]   Kirihara, T. Enhanced diffusion under irradiation. *J. Nucl. Mater.*, **1982**, *109*, 262-266.

[260]   Lajsan, V.B.; Schwartz, K.K.; Vitol, A.Ya. In: *Radiation Physics, III*; The effect of gamma-irradiation on the decay process of paramagnetic manganese centers in NaCl, "Zinatne": Riga, **1965**; pp. 103-110.

[261]   Watterich, A.; Voszka, R. Effect of X-irradiation on divalent metal aggregation in NaCl crystals at RT. *Phys. Stat. Sol. (b)*. **1979**, *93*, K161-165.

[262]   Opyrchal, H.; Nierzewski, K.D.; Macalik, B. Effect of γ-irradiation on Eu^{2+} ions in KCl crystals. *Phys. Stat. Sol. (b)*, **1982**, *112*, 429-434.

[263]   Opyrchal, H.; Nierzewski, K.D.; Drulis, H. Effect of γ-irradiation on EPR spectra of Eu^{2+} doped KCl and NaCl crystals. *Phys. Stat. Sol. (b)*, **1983**, *118*, K125-127.

[264]   Gotlib, V.I.; Trofimov, V.N.; Schwartz, K.K. Radiation-induced diffusion in KCl. *Izv. AN Latv. SSR, Fiz. i Tekhn. Nauk*, **1970**, N 6, 121-122.

[265]   Annenkov, Yu.M.; Galanov, Yu.I. In: *Spectroscopy of Dielectrics and Transfer Processes*; Radiation-stimulated diffusion in AHC, Leningrad, **1973**; pp. 81-85.

[266]   Zakhryapin, S.B.; Gladyshev, G.E.; Gromov, L.A. Diffusion of thallium in AHC under γ-irradiation. *Fiz. Tverd. Tela*, **1983**, *25*, 1152-1154.

[267]   Surzhikov, A.P.; Gyngazov, S.A.; Chernyavsky, A.V. In: *Materials and Articles thereof under the Influence of Various Types of Energy*. Investigation of foreign-valence impurity diffusion in KBr under intense electron irradiation, GUP "VIMI": Moscow, **1999**; pp. 113-115.

[268]   Indenbom, V.L. New hypothesis on the mechanism of radiation-induced processes. *Pis´ma v Zhurn. Techn. Fiz.*, **1979**, *5*, 489-492.

[269]   Aluker, E.D.; Gavrilov, V.V.; Indenbom, V.L.; Chernov, S.A. In: *Radiation Physics and Chemistry of Ionic Crystals*, Proceedings of the 5th Vsesoyuz. Sov.: Riga, Latvia, **1983**; pp. 42-44.

[270]   Koptelov, E.A. A probable scenario of copper precipitate clustering in model FeCu alloys under cascade-damage irradiation. *J. Nucl. Mater.*, **2007**, *366*, 238-247.

[271]   Kell, B.; Wollenberger, H. Radiated-enhanced formation rate of Guinier-Preston zones in proton-irradiated Cu-12.4at%Be. *J. Nucl. Mater.*, **1989**, *169*, 126-130.

[272]   Shriver, L. B.; Richardson, R.E. The effects of neutron irradiation on the microhardness of nickel and a nickel-carbon solid solution. *J. Nucl. Mater.*, **1982**, *108-109*, 451-455.

[273]   Sagaradze, V.V.; Ksitsyna, I.I.; Arbuzov, V.L.; Shabashov, V.A.; Filippov, Yu.I. Phase transformations in Fe-Cr alloys upon thermalaging and electron irradiation. *Fiz. Metal. i Metaloved.*, **2001**, *92*, N 5, 89-98.

[274]   Bystrov, L.N.; Ivanov, L.I.; Platov, Yu.M. Decomposition of supersaturated solid solution under eelectron irradiation. *Fiz. Metal. i Metaloved.*, **1968**, *25*, 950-953.

[275]   Poerschke, R.; Wollenberger, H. Interstitialcy diffusion in electron-irradiated Cu-Ni alloys. *Thin Solid Films*, **1975**, *25*, 167-170.

[276]   Repnikova, Ye.A.; Matvienko, I.A.; Chudinova, S.A.; Toropova, L.S.; Ustinovshchikov, V.M. Electron irradiation influence on precipitation of Al-Mg-Se alloy. *Fiz. Metal. i Metaloved.*, **1984**, *57*, 531-534.

[277]   Garcia, J.M.; Hernandez, J.A.; Murrieta, H.S.; Rubio, J.O. Effect of X-irradiation on impurity-vacancy dipoles in lead and calcium-doped NaCl. *Sol. Stat. Comm.*, **1983**, *47*, 515-518.

[278]   Mukherjee, M.L.; Capelletti, R. X-ray induced relaxation band due to higher aggregates in KCl-Pb. *Phys. Stat. Sol. (b)*, **1987**, *142*, 361-366.

[279]   Muccillo, R.; Rolfe, J. Effect of irradiation on impurity-vacancy dipoles in KBr crystals doped with strontium. *Phys. Stat. Sol. (b)*, **1974**, *61*, 579-587.

[280]   Gektin, A.V.; Cemushkova, V.I.; Charkina, T.A.; Shiran, N.V. Features of the impurity aggregation in KCl-Eu crystals. *Ukr. Fiz. Zhur.*, **1986**, *31*, 1232-1234.

[281]   Opyrchal, H.; Manfredi, M. Spectroscopic evidence of Eu-V dipole precipitation in KCl:Eu crystals. *Phys. Stat. Sol. (a)*, **1990**, *121*, 407-414.

[282]   Agrault, G. Radiation-induced precipitation in single- and dual-on irradiated Ti-6Al-4V. *J. Nucl. Mater.*, **1983**, *113*, 1-13.

[283]   Cauvin, R.; Martin, G. Radiation induced homogeneous precipitation in undersaturated solid-solutions. *J. Nucl. Mater.*, **1979**, *83*, 67-78.

[284]   Pareige, Ph.; Radiguet, B.; Barbu, A. Heterogeneous irradiation-induced copper precipitation in ferritic iron–copper model alloys. *J. Nucl. Mater.*, **2006**, *352*, 75-79.

[285]   Bakay, A.S.; Zelenskiy, V.F.; Matvienko, B.V.; Neklyudov, I.M.; Kolosov, I.E.; Parshin, A.M.; Orlov, A.N.; Trushin, Yu.V. Increased recombination of structural defects in the decomposition of solid solutions under irradiation. *VANT, Fiz. Rad. Povr. i Rad. Materialoved.* **1983**, *5(28)*, 3-11.

[286]   Rubio, O.J.; Munoz, F.A.; Patron, M. Manganese precipitation in NaCl induced by room temperature X-radiation. *Sol. Stat. Comm.*, **1985**, *55*, 109-112.

[287]   Aguilar, M.G.; Garcia Sole, J.; Murrieta, H.S.; Rubio, J.O. X-ray induced precipitation of Eu^{2+} in the alkali halides. *Rad. Eff.*, **1983**, *73*, 53-59.

[288]   Garcia, J.M.; Hernandez, J.A.; Murrieta, H.S.; Rubio, J.O. Effect of X-irradiation on impurity-vacancy dipoles in lead and calcium-doped NaCl. *Sol. Stat. Comm.*, **1983**, *47*, 515-518.

[289]   Cusso, F; Garsia Sole, J.; Murrieta, H.; Rubio, J.; Lopez, F.J. Luminescence spectra of Eu-impurity precipitated in the lattice of KI single crystals. Effects of X-irradiation on the precipitated phases. *Cryst. Latt. and Amorph. Mat.*, **1983**, *10*, 99-105.

[290]   Melikhov, V.D.; Roman'kov, S.E.; Volkova, T.V. Effect of ion irradiation on the aging processes in a Ti-48%Al-2%Nb alloy. *Fiz. Metal. i Metaloved.*, **2003**, *95*, N 2, 40-46.

[291]   Rusbridge, K.L. Dissolution of precipitates in Al-Ge during 200 keV Al^+ ion irradiation. *J. Nucl. Mater.*, **1983**, *119*, 41-50.

[292]   Potter, D.I.; McCormick, A.W. Irradiation-enhanced coarsening in Ni-12.8at.%Al. *Acta Metall.*, **1979**, *27*, 933-941.

[293]   Potter, D.I.; Wiedersich, H. Mechanisms and kinetics of precipitate restructuring during irradiation. *J. Nucl. Mater.*, **1979**, *83*, 208-213.

[294]   Sekimura, N; Zama, T.; Kawanishi, H.; Ishino, S. Precipitate stability in austenitic stainless steels during heavy ion irradiation. *J. Nucl. Mater.*, **1986**, *141-143*, 771-775.

[295]   Wollenberger, H. Atom transport mechanism in irradiation-driven phase transformation. *Mater. Sci. Forum*, **1987**, *15-18*, 1363-1378.

[296]   Wanderka, N.; Ramachandra, C.; Wahi, R.P.; Wollenberger, H. Radiation-altered phase stability of a precipitate-hardened copper alloy. *J. Nucl. Mater.*, **1992**, *189*, 9-13.

[297]   Plumton, D.L.; Kulcinski, G.L.; Dodd, R.A. Radiation induced precipitation in 9 MeV Al ion irradiated Ti-6Al-4V. *J. Nucl. Mater.*, **1987**, *144*, 264-274.

[298]   Baig, M.R.; Garawi, M.S. Investigation of movement of average solute concentration under irradiation environment. *Rad. Eff.*, **2004**, *159*, 203-208.

[299]   Potter, D.I.; Hoff, H.A. Irradiation effects on precipitation in γ/γ´ Ni-Al alloys. *Acta Metall.*, **1976**, *24*, 1155-1164.

[300]   Wagner, W.; Poerschke, R.; Axmann, A. Neutron-scattering studies of an electron-irradiated ^{62}Ni-41.4-at.%-^{65}Cu alloy. *Phys. Rev. B*, **1980**, *21*, 3087-3099.

[301]   Mukai, T.; Mitchell, T.E. Radiation-induced homogeneous precipitation in Ni-1at.%Be alloys. *J. Nucl. Mater.*, **1982**, *105*, 149-158.

[302]   Cawthorne, C.; Brown, C. The occurrence of an ordered fcc phase in neutron irradiated M316 stainless steel. *J. Nucl. Mater.*, **1977**, *66*, 201-202.

[303]   Ishino, S.; Chimi, Y.; Bagiyono; Tobita, T.; Ishikawa, N.; Suzuki, M.; Iwase, A. Radiation enhanced copper clustering processes in Fe–Cu alloys during electron and ion irradiations as measured by electrical resistivity. *J. Nucl. Mater.*, **2003**, *323*, 354-359.

[304]   Williams, T.M.; Titchmarsh, J.M.; Arkell, D.R. Void-swelling and precicpitation in a neutron-irradiated niobium-stabilised austenitic stainless steel. *J. Nucl. Mater.*, **1982**, *107*, 222-244.

[305]   Koptelov, E.A.; Subbotin, A.V. Reallocation of impurity atoms in a supersaturated solid solution by thermodiffusion in collision cascades. *J. Nucl. Mater.*, **2003**, *323*, 368-371.

[306]   Chee, S.W.; Stumphy, B.; Averback, N.Q. Vo, R.S.; Bellon, P. Dynamic self-organization in Cu alloys under ion irradiation. *Acta Mater.*, **2010**, *58*, 4088–4099.

[307]   Cauvin, R.; Martin, G. Solid solution under irradiation. II. Radiation-induced precipitation in AlZn undersaturated solid solutions. *Phys. Rev. B*, **1981**, *23*, 3333-3348.

[308]   Wilkes, P.; Liou, K.Y.; Lott, R.G. Comments on radiation induced phase instability. *Rad. Eff.*, **1976**, *29*, 249-251.

[309]   Yamauchi, H.; Sanches, J.M.; de Fontane, D.; Kikuchi, R.A. A thermodinamical approach to irradiation-induced precepitation in undersaturated solid solution. Gif-sur-Yevtte, CEA-DMECW, **1979**, 81-87.

[310]   Bocquet, J-L.; Martin, G. Irradiation-induced precipitation: a thermodynamical approach. *J. Nucl. Mater.*, **1979**, *83*, 186-199.

[311]   Cauvin, R.; Martin, G. Solid solution under irradiation. I. A model for radiation-induced metastability. *Phys. Rev. B*, **1981**, *23*, 3322-3332.

[312]   Urban, K.; Martin, G. Precipitate coarsening induced by point-defect recombination in alloys under irradiation. *Acta Met.*, **1982**, *30*, 1209-1218.

[313]   Bakay, A.S.; Turkin, A.A.; Turkin, Yu.A. Phase stability of binary alloys under irradiation. II. Radiation-modified phase diagrams. *Fiz. Metal. i Metalloved.*, **1991**, N 3, 77-85.

[314]   Gladyshev, G.E. The equilibrium concentration of impurities in the supersaturated solid solution during irradiation. *Zhur. Fiz. Khim.*, **1983**, *57*, 2872-2873.

[315]   Gladyshev, G.E. The effect radiation on the solid solution: thermodynamic approach. *Zhur. Fiz. Khim.*, **1982**, *56*, 1798-1800.

[316]   Gladyshev, G.E. The effect of radiation-defect formation on the impurity solubility in alkali halide crystals. *Zhur. Fiz. Khim.*, **1990**, *64*, 541-544.

[317]   Barbu, A.; Martin, G. Low-flux radiation-induced precipitation. *J. Appl. Phys.*, **1980**, *51*, 6192-6196.

[318]   Mruzik, M.R.; Russell, K.S. The effect of irradiation on the nucleation of incoherent precipitates. *J. Nucl. Mater.*, **1978**, *78*, 343-353.

[319]   Maydet, S.I.; Russell, K.C Precipitate stability under irradiation: point defect effects. *J. Nucl. Mater.*, **1977**, *64*, 101-114.

[320]   Nelson, .S.; Hudson, J.A.; Mazey, D.J. The stability of precipitates in an irradiation environment. *J. Nucl. Mater.*, **1972**, *44*, 318-330.

[321]   Bakay, A.S.; Kiryukhin, N.M. On the evolution of precipitates in aged alloy under irradiation. *VANT, Fiz. Rad. Povr. i Rad. Materialoved.* **1983**, *5(28)*, 33-40.

[322]   Bakay, A.S.; Voevodin, V.N.; Zelenski, V.F.; Kiryukhin, N.M.; Matvienko, B.V.; Nekludov, I.M.; Platonov, P.V. Variation of precipitation size distribution during irradiation of nimonik-type alloy by heavy ions. *Fiz. Metal. i Metaloved.*, **1988**, *66*, 619-621.

[323]   Abyzov, A.S.; Slezov, V.V.; Tanatarov, L.V. Irradiation-induced growth of new phase inclusions. *Fiz. Tverd. Tela*, **1991**, *33*, 834-839.

[324]   Krishan, K.; Abromeit, C. Calculation of radiation-induced instability in concentrated alloys. *J. Phys. F*, **1984**, *14*, 1103-1116.

[325]   Wagner, W.; Poerschke, P.; Wollenberger, H. Short-range clustering and long-range periodic decomposition of an electron irradiated Ni-Cu alloy. *J. Phys. F*, **1982**, *12*, 405-424.

[326]   Gladyshev, G.E.; Shachetova, E.S. Radiation-induced decomposition of solid solutions on the based alkali halide crystals. *Zhur. Fiz. Khim.*, **1996**, *70*, 162-163.

[327]   Nikolayev, V.A.; Zhukov, O.N; Shapovalov, S.V. Non-equilibrium segregation and damage of bcc Fe-Ni alloys. *Fiz. Metal. i Metalloved.*,**1989**, *68*, 578-581.

[328]   Anthony, T.R. In: *Radiation-induced voids in metals and alloys*, J.W. Corbett and L.C. Ianniello, Eds; AEC Symp. Series, Conf.-701601, **1972**, 630.

[329]   Geits, S.F.; Natsvlishvili, G.I. In: *Electronic and Ionic Processes in Solids, VIII;* Radiation changes in the microstructure of lithium fluoride, Metsniereba: Tbilisi, **1975**; pp. 47-51.

[330]   Turos, A.; Meyer, O. Impurity-defect interaction in ion-implanted supersaturated Au-Fe alloys. *Phys. Rev. B*, **1985**, *31*, 5694-5702.

[331]   Clausing, R.E.; Heatherly, L.; Faulkner, R.G.; Rowcliffe, A.F.; Farrell, K. Radiation-induced segregation in HT-9 martesitic steel. *J. Nucl. Mater.*, **1984**, *126*, 978-981.

[332]   Zhang, Y.G.; Jones, I.P. Electron irradiation of aluminum-zinc alloys: 1. Radiation-induced segregation in an aluminum-0.35at% zinc alloy. *J. Nucl. Mater.*, **1989**, *165*, 252-265.

[333]   Zhang, Y.G.; Jones, I.P. Electron irradiation of aluminum-zinc alloys: 3. Radiation-enhanced precipitation in an aluminum-4.5at% zinc alloy. *J. Nucl. Mater.*, **1989**, *165*, 278-285.

[334]   Nakata, K; Masaoka, I. Solute segregation along non-migrated and migrated grain boundaries during electron irradiation in austenitic stainless steels. *J. Nucl. Mater.*, **1987**, *150*, 186-193.

[335]   Agarwal, S.C.; Rehn, L.E.; Nolfi, F.V. Irradiation-induced void swelling and solute segregation in a V-ion-irradiated V-15wt%Cr bcc alloy. *J. Nucl. Mater.*, **1978**, *78*, 336-342.

[336]   Wang, Z.; Ayrault, G.; Wiedersich, H. Segregation in irradiated titanium alloys. *J. Nucl. Mater.*, **1982**, *108-109*, 331-338.

[337]   Allen, T.R.; Cole, J.I.; Gan, J.; Was, G.S.; Dropek, R.; Kenik, E.A. Swelling and radiation-induced segregation in austenitic alloys. *J. Nucl. Mater.*, **2005**, *342*, 90-100.

[338]   Ashworth, M.A.; Norris, D.I.R.; Jones, I.P. Radiation-induced segregation in Fe-20Cr-25Ni-Nb based austenitic stainless steels. *J. Nucl. Mater.*, **1992**, *189*, 289-302.

[339]   Hautojarvi, P.; Pollanen, L.; Vehanen, A.; Yli-Kauppila, J. Vacancies and carbon impurities in α-iron: neutron irradiation. *J. Nucl. Mater.*, **1983**, *114*, 250-259.

[340]   Kornblit, L.; Ignatiev, A. The size effect in radiation-induced segregation of solutes in binary metallic alloys. *J. Nucl. Mater.*, **1984**, *126*, 77-78.

[341]   Okamoto, F.R.; Rehn, R.E. Radiation-induced segregation in binary and ternary alloys. *J. Nucl. Mater.*, **1979**, *83*, 2-23.

[342]   Hackett, M.J.; Busby, J.T.; Miller, M.K.; Was, G.S. Effects of oversized solutes on radiation-induced segregation in austenitic stainless steels. *J. Nucl. Mater.*, **2009**, *389*, 265-278.

[343]   Wang, Z.; Aurault, G.; Wiedersich, H. Radiation-induced segregation in titanium alloys. *J. Nucl. Mater.*, **1983**, 118, 109-114.

[344]   Lu, Z.; Faulkner, R.G.; Sakaguchi, N.; Kinoshita, H.; Takahashi, H.; Flewitt, P.E.J. Control of phosphorus inter-granular segregation in ferritic steels. *J. Nucl. Mater.*, **2004**, *329-333*, 1017-1021.

[345]   Kato, T.; Takahashi, H.; Izumija, M. Grain boundary segregation under electron irradiation in austenitic stainless modified with oversized elements. *J. Nucl. Mater.*, **1992**, *189*, 167-174.

[346]   Yoshida, Y.; Fratzl, P.; Volg, G.; Höfer, H.; Dworschak, F. Radiation-iduced segregation in proton irradiated <u>Au</u>Fe studied by Mössbauer spectroscopy. *J. Phys.: Condens. Matter*, **1992**, *4*, 2415-2428.

[347]   Janghorban, K.; Ardell, A.J Irradiation damage in proton irradiated palladium-iron solid solution. . *J. Nucl. Mater.*, **1983**, 114, 66-74.

[348]   Brimhall, J.L.; Baer, D.R.; Jones, R.H. Radiation induced phosphorus segregation in austenitic and ferritic alloys. *J. Nucl. Mater.*, **1984**, 122, 196-200.

[349]   Arbuzov, V.L.; Danilov, S.Ye.; Klotsman, S.M.; Tatarinova, G.; Timofeyev, A.N. Formation of sulphur-pure zones in vicinity of nickel surface during irradiation. *Fiz. Metal. i Metaloved.*, **1988**, *66*, 610-612.

[350]   Erck, R.A.; Rehn, L.E. Kinetics of radiation-induced segregation in molybdenum-rhenium alloys. *J. Nucl. Mater.*, **1989**, *168*, 208-219.

[351]   Kark, G.S.; Astaf'yev, A.A.; Markov, S.I. Connection between radiation embrittlement and temper brittleness of low-alloyed steel. *Fiz. Metal. i Metaloved.*, **1984**, *57*, 592-598.

[352]   Watanabe, H.; Muroga, T.; Yoshida, N.; Kitajima, K. Precipitate resolution in an electron irradiated Ni-Si alloy. *J. Nucl. Mater.*, **1988**, *158*, 179-187.

[353]   Yin, Y.F.; Faulkner, R.G.; Lu, Z. Irradiation-induced precipitation modelling of ferritic steels. *J. Nucl. Mater.*, **2009**, *389*, 225-232.

[354]   Pechenkin, V.A.; Epov, G.A. Theoretical studies of radiation stimulated segregation of components of binary and ternary substutional alloys on point defect outlets. *Fiz. Metal. i Metalloved.*, **1991**, N 3, 58-66.

[355]   Lam, N.Q.; Janghorban, K.; Ardell, A.J. On the modeling of irradiation-induced homogeneous precipitation in proton-bombarded Ni-Si solid solutions. *J. Nucl. Mater.*, **1981**, *101*, 314-325.

[356]   Johnson, R.A.; Lam, N.Q. Solute segregation in metals under irradiation. *Phys. Rev. B*, **1976**, *13*, 4364-4375.

[357]   Johnson, R.A.; Lam, N.Q. Solute segregation under irradiation. *J. Nucl. Mater.*, **1978**, *69/70*, 424-433.

[358]   Murphy, S.M. A model for segregation in dilute alloys during irradiation. *J. Nucl. Mater.*, **1989**, *168*, 31-42.

[359]   Hobbs, J.E.; Marwick, A.D. Measurements of radiation-induced segregation and diffusion in an ion irradiated alloy using implanted manganese. *Nucl. Inst. and Meth. B*, *9*, 169-177.

[360]   Bartels, A.; Dworschak, F.; Weigert, M. Kinetics of radiation-induced segregation in electron-irradiated dilute NiSi and NiGe alloys. *J. Nucl. Mater.*, **1988**, *152*, 82-93.

[361]   Lam, N.Q.; Janghorban, K.; Ardell, A.J. On the modeling of irradiation-induced homogeneous precipitation in proton-bombarded Ni-Si solid solutions. *J. Nucl. Mater.*, **1981**, *101*, 314-325.

[362]   Okamoto, P.R.; Wiedersich, H. Segregation of alloying elements to free surfaces during irradiation. *J. Nucl. Mater.*, **1974**, *53*, 336-345.

[363]   Lam, N.Q.; Okamoto, P.R.; Johnson, R.A. Solute segregation and precipitation under heavy-ion bombardment. *J. Nucl. Mater.*, **1978**, *78*, 408-418.

[364]   Sorokin, M.V.; Ryazanov, A.I. Effect of elastic stress field near grain boundaries on the radiation induced segregation in binary alloys. *J. Nucl. Mater.*, **2006**, *357*, 82-87.

[365]   El-Ashram, T.; Asaad, Y. Radiation annealing of rapidly solidified Sn-6.7Sb-5.3Zn alloy by low-dose gamma ray. *Rad. Eff.*, **2008**, *163*, 843-849.

[366]   Adam, J.; Green, A.; Dugdale, R.A. An effect of electron bombardment on order in *Cu₃Au* alloy. *Phil. Mag.*, **1952**, *43*, 1216-1218.

[367]   Gilbert, J.; Herman, H., Damask, A.C. Electron irradiation of Cu₃Au. *Rad. Eff.*, **1973**, *20*, 37-42.

[368]   Decker, D.; Dworschak, F.; Lehmann, C.; Rie, K.T.; Schuster, H.; Wollenberger, H.; Wurm, J. Determination of the average number of replacement collisions in an electron-irradiated f.c.c. crystal (Ni₃Mn). *Phys. Stat. Sol.*, **1968**, *30*, 219-229.

[369]   Koczak, M.J.; Herman, H.; Damask, A.C. The production and annealing of point defects in β-CuZn. *Acta. Met.*, **1971**, *19*, 303-310.

[370]   Blewitt, T.H.; Coltman, R.R. Radiation ordering in Cu₃Au. *Acta Met.*, **1954**, *2*, 549-551.

[371]   Blewitt, T.H.; Coltman, R. The effect of neutron irradiation on metallic diffusion. *Phys. Rev.*, **1952**, *85*, 384-384.

[372]   Halbwachs, M.; Hillairet, J. Assignment of radiation-enhanced ordering to vacancies and self-interstitials, in α-AgZn alloys. *J. Phys. F*, **1981**, *11*, 2247-2256.

[373]   Siegel, S. Effect of neutron bombardment on order in the alloy Cu₃Au. *Phys. Rev.*, **1949**, *75*, 1823-1824.

[374]   Howe, L.M.; Rainville, M.H. A study of the irradiation behaviour of Zr₃Al. *J. Nucl. Mater.*, **1977**, *68*, 215-234.

[375]   Schulson, E.M. The ordering and disordering of solid solutions under irradiation. *J. Nucl. Mater.*, **1979**, *83*, 239-264.

[376]   Carpenter, G.J.C.; Schulson, E.M. The disordering of Zr₃Al by 1 MeV electron irradiation. *J. Nucl. Mater.*, **1978**, *73*, 180-189.

[377]   Kirk, M.A.; Blewitt, T.H.; Scott, T.L. Irradiation disordering of Ni₃Mn by replacement collision sequences. *Phys. Rev. B*, **1977**, *15*, 2914-2922.

[378]   Banerjee, S.; Urban, K. Kinetics of order-disorder transformation in alloys under electron irradiation. *Phys. Stat. Sol. (a)*, **1984**, *81*, 145-162.

[379]   Banerjee, S.; Urban, K.; Wilkens, M. Order-disorder transformation in Ni₄Mo under electron irradiation in a high-voltage electron microscope. *Acta Metall.*, **1984**, *32*, 299-311.

[380]   Holz, M.; Ziemann, P.; Buckel, W. Direct evidence for amorphization of pure gallium by low-temperature ion irradiation. *Phys. Rev. Lett.*, **1983**, *51*, 1584-1587.

[381]   Aronin, L.R. Radiation damage effects on order-disorder in nickel-manganese alloys. *J. Appl. Phys.*, **1954**, *25*, 344-349.

[382]   Jaouen, C.; Delafond, J.; Riviere, J.P. Crystalline to amorphous transformation in NiAl: ion irradiation studies in relation to cascade parameters. *J. Phys. F*, **1987**, *17*, 335-350.

[383]   Thome, L.; Benyagoub, A.; Audouard, A.; Chaumont, J. Amorphous Fe-B alloys produced by ion implantation. I: Electrical properties. *J. Phys. F*, **1985**, *15*, 1237-1247.

[384]   Matsukawa, Y.; Suda, T.; Ohnuki, S.; Namba, C. Microstructure and mechanical properties of neutron irradiated TiNi shape memory alloy. *J. Nucl. Mater.*, **1999**, *271-272*, 108-110.

[385]   Parsons, J.R. Conversion of crystalline germanium to amorphous germanium by ion bombardment. *Phil. Mag.*, **1965**, *12*, 1159-1178.

[386]   Naguib, H.M.; Roger Kelly. Criteria for bombardment-induced structural changes in non-metallic solids. *Rad. Eff.*, **1975**, *25*, 1-12.
[387]   Brimhall, J.L.; Kissinger, H.E.; Charlot, L.A. Amorphous phase formation in irradiated intermetallic compounds. *Rad. Eff.*, **1983**, *77*, 237-293.
[388]   Swanson, M.L.; Parsons, J.R.; Hoelke, C.W. Damaged regions in neutron-irradiated and ion-bombarded Ge and Si. *Rad. Eff.*, **1971**, *9*, 249-256.
[389]   Lu, Z.; Faulkner, R.G.; Morgan, T.S. Formation of austenite in high Cr ferritic/martensitic steels by high fluence neutron irradiation. *J. Nucl. Mater.*, **2008**, *382*, 223-228.

Index

www.ingramcontent.com/pod-product-compliance
Lightning Source LLC
Chambersburg PA
CBHW041720210326

41598CB00007B/719